First to Die

About the cover

A chaotic scene on the deck of the sinking ship *Vestris*

This dramatic photograph, shown on the cover, appeared on the front page of the New York *Daily News* on Thursday, November 15, 1928. It was taken about noon on Monday, November 12, 1928, by Fred Hanson, a pastry chef on the *Vestris*, using a Kodak folding camera that he bought for $8.50 the day before the ship sailed. ($8.50 in 1928 would be about $118.00 in 2015 dollars.) The people are trying to get to the lifeboats on the port side of the ship. This picture has been rotated from the original version to show the actual angle of the ship's list, which was very close to 29 degrees. Steward George Hogg (far left) is not praying as some have said; both of his arms were broken and he could do nothing to help.

"Women and children were first—to drown."
—"Catastrophe: *Vestris*,"
Time, November 26, 1928

First to Die
The Tragic Sinking of the SS *Vestris*

G. David Thayer & Kristin Delaplane

With a Foreword by Ramon Jackson
Deep–Ocean Bathymetrist

This book was written and typeset with Microsoft Publisher 2010 and printed to PDF using Adobe Acrobat X.

Printed on 60-pound acid-free archival paper for longevity.
Made in the United States of America.

Published jointly by Rapidsoft Press®, Sarasota, Florida (SAN 299-5840), and Our American Stories® LLC, Tucson, Arizona.
www.rapidsoftpress.com & www.ouramericanstories.com

ISBN 978-0-9663909-7-1

ACKNOWLEDGMENTS

This book would not have been possible without the splendid coverage given to the sinking of the *Vestris* by the newspapers of the day, particularly in November and December of 1928. Grateful appreciation is due especially to the *New York Times*, *Time* magazine, and the photo archives department of the New York *Daily News*. We must also acknowledge the contributions of those whose families were impacted by this tragedy.

A special thank you goes out to Ramon Jackson, who maintains a website for the estate of his father, Judson G. Jackson, whose parents and brother were lost in the disaster. He has assembled a large collection of links to documents and stories about the *Vestris*, which sank along with most of her passengers and much of her crew on November 12, 1928. See http://patriot.net/~eastlnd2/rj/vestris

The New Twin-Screw Steamers "Vestris" and "Vandyck."

Interesting Features of the Fleet.

The largest and finest vessels engaged in North and South Atlantic transportation are those of the Lamport & Holt Line, leaving New York every two weeks. Their adaptation to the service in which they are engaged is nothing short of perfect.

The conditions prevailing on the southern route are different from those attending the ordinary transatlantic course. One passes from zone to zone with varying climates and temperatures. In order to equalize these conditions as far as possible and maintain a maximum of comfort, it is necessary that the vessels should be specially constructed. And that is one consideration that gives pre-eminence to the steamers of this fleet; they were built with a sole view to this particular service.

One important feature is the perfect ventilation secured by modern appliances. The result is that a delightful uniformity of pure atmosphere is maintained all the way between New York and Argentina. This comfort is increased by a sufficient number of electric fans. Travelers find that the pure sea air in the dining saloon is conducive to appetite, and the cuisine is not only kept up to the most creditable standard, but is adapted to the varying climates as they are reached from stage to stage. An unusual number of outside staterooms are provided, and all of the staterooms are notably homelike and comfort-

A page from the Lamport & Holt Line brochure, ca 1912

TABLE OF CONTENTS

Source note methodology

Sources for the material in each chapter of this book are contained in a series of endnotes. These endnotes are numbered sequentially for each chapter, beginning with the number 1. Citations in the text of each chapter are given by superscripts that give the number of the endnote for that particular source. These superscript numbers are not necessarily in numerical order within each chapter, although the first instance of each such citation does appear in numerical order. Hence, the superscript [12], for example, refers to Endnote 12 of the chapter in which it appears, regardless of whether or not a preceding superscript may have a higher number. In other words, the endnote numbers may not always be in numerical order within each chapter. But the superscript [12] will always mean Endnote 12 for the chapter in which it appears, no matter where it is located in that chapter.

In a number of places we have long sections of text, formatted in different ways (body text, quotations), but all taken from the same source. For these long sections, we place a superscript section mark [§] at the beginning. All of the material from there up to the next superscript number comes from the source in that endnote.

A NOTE TO OUR READERS

Many of the documents and newspaper articles quoted in this book use the term "Negro" to refer to Blacks. That was 1928, not 2015, and this was common usage back then. We decided not to redact these instances. Most of the Blacks were West Indian, so the term "African American" does not apply to them.

Foreword

The loss of the *Vestris* has always been a significant family event. My grandfather, Dr. Ernest Alonzo Jackson, his wife, Jannette Beazley Jackson, and their youngest son Cary were all lost when the *Vestris* sank on November 12, 1928. They had left behind the older brothers and sisters of Cary Jackson, who were either in school or starting careers in the United States. One of the sons left behind was my father, Judson Gordon Jackson, who compiled some files on the sinking of the *Vestris*, including letters from survivors that he had requested. There were some family mementos saved because of baggage that failed to make the sailing. Though of old Virginia lineage, the family had been based in Brazil because my grandfather had gone there as a missionary just before the turn of the century.

I took another look at my father's *Vestris* files after retiring from a working life that took me to sea frequently over a period of twenty years. I was struck anew at the way that story is an example of how to lose a ship and many lives.

Some examples of how such a thing can happen:

- Corporate masters concentrating on finance dictate how seamen, charged with safety of a vessel that may already be suffering from cost-cutting neglect, handle emergencies.

- The owners cover up negligence in enforcing regulations by those charged with inspections.

- Poor to downright bad leadership allows gaps in vigilance and information passed from crew to officers.

In a wider arena, the *Vestris* sinking is an example of how misinformation becomes "fact." The phrase "lost due to a storm" originates from ignorant news reporters, or perhaps from calculated disinformation, and has made its way into some popular modern references. Although the sinking of the *Vestris* is less well known than that of the *Titanic*, they have in common neglect in heeding warnings and ship's masters who seem amazingly passive. To me these are hints of the attitudes of an age where perhaps a captain's competence and ability to act was heavily subordinated to social class and the ability to get along with the gentlemen among the passengers.

None of these interesting aspects have been reviewed in decades so far as I know. *First to Die* exposes aspects of the *Vestris* case that are significant in correcting past misinformation. In particular, the positional information is a new analysis of facts that became associated with the sinking through popular news accounts and worked their way into court documents without rebuttal.

One fact is the alleged position error of the SOS that is shown in this research to likely not be an error at all. Among all the failings of that ship's owners, captain, and company, a blunder in the ship's position as given in the SOS does not appear to be among them. Ship positions far from land, prior to our modern systems, had fairly large inaccuracies, a good one being within a couple of miles of actual position. In the case of the *Vestris*, we find that not only did that SOS position have the benefit of radio direction finding data from shore, but the error was likely made by those thinking that the lifeboats and debris field would still be at the original position of the sinking many hours later.

Today any naval- or coast-guard-type rescue effort by any technological nation will include oceanographic and weather data for predicting where and how debris fields, including the people and bodies, will move from a disaster point over time. In the rescue effort mounted to save the survivors of the *Vestris*, it appears only one rescue ship's master did that while everyone from news organizations to court experts assumed that one starts looking at the place where the disaster happened without due regard to where and how things drift at sea over hours or days.

Modern examples of corporate greed, shortsightedness, inattention to enforcement of regulations, and lack of professional conduct make a new look at the sinking of the *Vestris* worthwhile.

The *Vestris* disaster did lead to some corrective actions in law and regulation. It served as an example to the public of how a sorry state of professional and material condition aboard a ship makes safety seem like a joke. It is a lesson that apparently has to be learned again and again. Witness big ships recently rolling on their sides after questionable actions—such as the loss of the *Costa Concordia* in 2012. At the least the captain of the *Vestris* was a gentleman who did go down with his ship.

—Ramon Jackson
Fairfax, Virginia
12 November 2015

Some passengers from the *Vestris*

Clockwise from top left: Marion Batten of Altoona, Pennsylvania, who was saved; her husband, Norman K. Batten, a race car driver, who was lost; Captain William Sears of Brooklyn, who was saved; William P. Adams of Chicago and Odebolt, Iowa, who was saved; Vincenzo Murri of Philadelphia, who was lost; Gaetano Abbadini of Philadelphia, who was saved; Gladys Stevens of Buenos Aires, Argentina, who was lost; and Orrin S. Stevens, husband of Gladys, who was saved. The given names of Abbadini and Murri were misspelled in the captions above. Photos reprinted from the *New York Times* for November 14, 1928.

Preface

THIS IS THE STORY OF THE LAST DAYS of the steamship *Vestris*. She was built in Ireland in 1912 by master shipbuilders for the Lamport & Holt Line. The construction method used for her was the latest and most advanced of its day. Lamport & Holt wrote in their brochure that "the magnificent SS *Vestris* was named after a family whose exquisite dancing many years ago won enthusiastic plaudits from all the stages of Europe." She was given a rating of "100 A1"—the highest—by Lloyd's Register, a maritime agency that provided risk assessment and ship certification services.

The *Vestris* weighed 10,660 gross tons, had twin screw propulsion, a speed of 15 knots, and could carry 280 first-class, 130 second-class, and 200 third-class passengers with a crew of 250. Launched on May 16, 1912, she made her maiden voyage on September 19, 1912.

The *Vestris* was a shelter deck vessel (an oceangoing vessel with a continuous top deck above the main deck), having three complete decks extending from stem to stern, namely shelter deck, upper deck, and main deck; there was also a lower deck extending from her stem to the aft end of No. 2 cargo hold forward. On the shelter deck was a cross alleyway extending the width of the ship. On the starboard side leading forward from the cross alleyway was the firemen's passage or alleyway. The forward part of the shelter deck was an open deck generally known as a "well deck." The decks and lifeboats can be seen in the photograph on the facing page.

The *Vestris* carried fourteen lifeboats. Four were carried on the poop deck (a deck that forms the roof of a cabin built in the rear, or "aft," part of a ship's superstructure). Ten were carried under davits on the top deck. The five on the port side were given even numbers, those on the starboard side, odd numbers.

For years, the *Vestris* sailed on the North and South Atlantic oceans successfully. During the Great War, a German torpedo just missed her. Later, she suffered a fire in her coal bunkers that took fourteen days to extinguish. In 1922, she began plying the New York to Buenos Aires route, a voyage she completed many times.

All of this success came to a tragic halt during three days in November of 1928. A virtual confluence of calamities descended upon the *Vestris* like a great plague. It began with a botched pre-voyage inspection that failed to reveal the truth: that the *Vestris* was a decrepit vessel made so by continual abuse and neglect. Everything from lifeboats to port covers was defective. So long as things went well this was not a problem, but this time they did not go well.

Within hours of departing from the port of Hoboken, New Jersey, water began leaking into the ship through defective ports, which were too close to the waterline. The *Vestris* was sailing with her salt water draft some eight inches below the required level. These leaks grew worse overnight. She was battered by a powerful nor'easter that hit during the night, and water poured into the ship so fast that by noon the next day she had a list of some seven degrees to starboard.

The list grew steadily worse. By Sunday evening, it was about twelve degrees, and by Monday morning, it was almost twenty degrees. Just before the ship capsized, her list was roughly forty degrees.

The SS *Vestris* about 1912
The *Vestris* in her prime. Photo taken from a Lamport & Holt brochure.

The beleaguered captain of the ship, William John Carey, was caught between a rock and a hard place. On the one hand, the Lamport & Holt instructions were to avoid salvage if possible; on the other hand, his passengers and crew were in desperate straits.

The ship's pumps could not handle the deluge of water that was coming into the *Vestris*, yet Chief Engineer James Avard Adams kept telling Captain Carey that the ship could still be saved. When it became apparent that she was going to sink, the captain gave the order to send out an SOS. It was too late.

At 2:30 P.M. on Monday, November 12, 1928, in a cacophony of children screaming, women wailing, men cursing, and two lifeboats clanking against the sides of a ship they would never leave, the *Vestris* rolled over "like a big potato in a tub of water," as one of the survivors put it, and sank to the bottom, some 215 statute miles southeast of Atlantic City, New Jersey.

The story of the SS *Vestris* was over. Fully 111 persons—68 passengers and 43 of the ship's officers and crew—lost their lives. But the stories of many of the 214 survivors remain to be told here.

Brochure cover for the Lamport & Holt Line, ca 1912

For the voyage of the *Vestris* beginning on November 10, 1928, she was to follow the inner route, with a port of call at Barbados, British West Indies.

Chapter One
The Saga Begins

SECOND STEWARD ALFRED DUNCAN LOOKED HIS CAPTAIN in the eye and yelled, "You'd better jump, sir!" "Hell no!" replied Captain William Carey. "You jump!"

The captain, looking worn and haggard, was standing on the bridge of his sinking ship wearing his heavy sea coat and no life belt. He muttered, "My God, my God, I am not to blame for this," as the water rose around him. He yelled to Second Officer Leslie Watson, "Jump, Watson!" But there wasn't time. The SS *Vestris* sank beneath the waves before any of them could make a move. Of these three, only Duncan and Watson—both wearing life belts—would live to tell this tale. True to his calling, Captain William John Carey went down with his ship.

~ ~

WHEN SHE WAS BUILT IN 1912, the *Vestris* was classed 100 A1—the highest possible grade—by Lloyd's Register, a British maritime ship certifier (not connected with the famous Lloyd's of London insurance services). Before her last voyage in 1928, the British

Lamport & Holt ship had been overhauled in dry dock in Brooklyn, New York. She was then examined over three days (November 5–7, 1928) by three inspectors from the US Steamboat Inspection Service. One inspector admitted later that the *Vestris* listed slightly at her dock while she was being inspected. The inspector said that every lifeboat was tested: filled with men, lowered to the water, and raised again. This statement was discovered in later testimony to have been a lie. Nonetheless, the *Vestris* was certified as seaworthy and equipped according to law. As she was being escorted by tugboats to a pier in Hoboken, New Jersey, on November 9, 1928, a minor collision in the Erie Basin allegedly did no more than "scrape paint." The description of this collision as "minor" was called into question during later investigations of the sinking.

From the Hoboken pier, the *Vestris* started her final voyage bound for Buenos Aires, with a port of call at Barbados. She had been sailing the New York to South America route for sixteen years. Her fatal voyage began at 3:45 P.M. on Saturday, November 10, 1928.

Some passengers had concerns

Prior to the passengers boarding the ship, several of them did take note of a list to starboard (a leaning to the right side) of the vessel. After the fact, passenger Walter Spitz blamed improper loading of the ship's cargo. Spitz, of Berlin, Germany, was on a pleasure trip to South America. In an interview he stated:

> I was on the pier at a quarter of two on sailing day and standing with Engineer [Fred W.] Puppe—another passenger— and we were both surprised to see them loading all the cargo on one side. I understood from some other people that in such cases the cargo is straightened out as soon as the ship reaches the open sea but in our case there was no chance to do so as we had stormy weather from the start. In ships I have traveled on the world over, the straightening out of the cargo is done before the ship sails.[1]

Forty-five-year-old Fred W. Puppe, a mechanical engineer, was traveling with his wife and seven-month-old daughter to take an executive position in Buenos Aires. He echoed his friend's concern:

> I am an engineer employed by the Standard Oil Company and am used to overseeing such matters as the loading of cargo. This wreck was due to deliberate criminal procedure. The ship was doomed and lost Sunday morning, thirty-six hours before she went down. The disaster was caused by deliberately bad loading of the cargo, which shifted a few hours after we left port and couldn't be rearranged. All the portholes began to leak and take in water as soon as the ship began leaning over at a slight angle. Everything about the ship was in bad condition.[2]

First-class passenger William Carlos Quiros of Washington, DC, chancellor to the Argentine consulate in New York, was bitter in his criticism of the way the *Vestris* was handled, writing:

> She had a list when tied up at the pier before sailing. . . . Two friends, both of them women, warned me that the *Vestris* had a list. "You will have to walk on one side," they said.... We noticed it in walking around the deck. However, I thought nothing of it then. I don't think any of us did. I had met most of the passengers as I had given them their passports in the consulate. We chatted among ourselves without the slightest thought to danger.[3]

Orrin S. Stevens, a representative in Buenos Aires of the First National Bank of Boston, said that when he and his wife walked onto the deck of the *Vestris*, "I noticed the list, but I thought to myself, *This is a British ship—it must be safe*. If it had been of another nationality I wouldn't have sailed on it."[4]

Edward M. Walcott, a Georgetown, British Guiana, shipping agent, stated that when he came aboard the *Vestris*, he "noticed a slight list, but nothing to worry about."[5]

At a later inquiry, a waiter from the *Vestris* testified that in his opinion she had always had a "barely noticeable" list.[6]

A botched inspection doomed the *Vestris*

Although everything seemed in order when the *Vestris* was inspected prior to her voyage slated for November 10, 1928, trouble lurked behind those inspection reports. The official inspection form (shown opposite) indicated that the lifeboats and other safety equipment had been inspected and found to be in good condition. But as it would turn out later, the steamboat inspector, Edward Keane, lied when he said he had inspected the lifeboats as required. The report was signed by J. L. Crone, supervising inspector.

Each lifeboat was to have been filled with men to capacity, lowered to the water, and then pulled back up by the inspectors. They did not do this. Not only that, but they gave the lifeboats and their davits and lowering equipment only a cursory inspection. When the crew of the *Vestris* attempted to lower the lifeboats as the ship was sinking, they found the equipment was rusty and often did not work. Some lifeboats would not release when lowered, and the davit ropes had to be cut with knives to free them.

And remember the "minor" collision in the Erie Basin as she left dry dock that "did no more than scrape paint"? There was more to it than that. An unnamed British expert who had been associated with many Board of Trade inquiries into sea disasters was quoted in English newspapers as saying:

> On the night she left New York, the *Vestris* was in a collision in New York Harbor with the [Grace Line] steamer *Santa Luisa*. A photograph has appeared showing the *Santa Luisa* with her bow in collision with the starboard side of the *Vestris*, and all evidence points to . . . the *Vestris* [having] had a heavy list to the starboard before the sinking. Was there any connection between this collision and the sinking of the *Vestris*? Why has no evidence from the *Santa Luisa* been called in New York?[7]

There is some controversy about the details of this incident. The testimony that Charles Verchere, assistant radio operator on the

INSPECTION CERTIFICATE UNDER WHICH VESTRIS CLEARED ON FATAL VOYAGE.

The Certificate of Inspection for the *Vestris*

This certificate, as reproduced by the *New York Times* edition of November 16, 1928, shows that Supervising Inspector J. L. Crone signed off on all of the safety equipment aboard the *Vestris*, including the lifeboats, but Assistant Inspector of Hulls Edward Keane inspected the hull and lifeboats.

Vestris, gave on November 16, 1928, while being examined by Attorney Charles H. Tuttle, was that the collision with a ship "that looked like a cargo boat" occurred on the port beam of the *Vestris* and that it "could hardly be called a collision."

But a photograph of this collision, which appeared on page two of the *Brooklyn* (New York) *Daily Eagle* of November 18, 1928, trumps testimony by a possibly biased witness, so it certainly seems plausible that this minor collision could have damaged or loosened some of the plates on the starboard side of the *Vestris* even before she started out on her fatal voyage. Any moving contact between two multi-thousand-ton ships is a serious matter.

The ship was overloaded

The *Vestris* was loaded with cargo, coal, water, stores, baggage, mail, permanent ballast weights, and water in her ballast tanks, variously estimated at between 7,370 and 7,665 tons. The cargo consisted mostly of heavy machines in large cases and 1,097 sacks of mail, including diplomatic correspondence with US consular agents. She carried 4,612 automobile bodies—the greater number being truck bodies—and accessories, according to the manifest of the cargo filed with the customs authorities. Most of the machines were being sent to Buenos Aires; a few were consigned to other ports. There were also 2,000 bags of cement.

More cargo included fresh fruits and textiles, including cotton goods and piece goods. Two cases of revolvers, thirty-one of cartridges, and several cases of children's toys were also included. Other goods were insecticides, toy balloons, safety razors, phonographs and records, building and other machinery, office equipment—including adding machines, cigarettes, cigars, lampshades, gym equipment, roller skates, and one case of soda straws.

This proved to be more cargo than the *Vestris* could safely carry in the event of a storm.

The voyage begins

At 3:45 P.M. on November 10, 1928, tugboats pulled the *Vestris* away from her Hoboken pier.

First-class passenger Fred Puppe went to his cabin only to discover that most of his baggage was missing. He finally found it in the hold, where it had been tossed with the heavy baggage. Later he learned that other passengers had experienced similar problems with their luggage. "The handling of the baggage was absolutely careless," Puppe declared. He asked a steward for help and was told: "I'm not supposed to take care of your cabin, but I will do the best I can to get around to you because your steward is too drunk and will be unable to attend to you."[8]

Puppe's description of conditions at the start of the voyage hardly amounts to that of an auspicious beginning.

Signs of the disaster start soon

Trouble began to show up as early as the first day out of port.

Puppe recalled that first night at sea: "At dinner Saturday night we remarked at the speed the ship was making. We said to ourselves, 'If it goes like this all along we'll be there in no time.'"[9] Some passengers, however, noted the absence of the captain at dinner that first night. An article in *Time* magazine stated: "Saturday evening, Captain Carey did not appear in the dining room for dinner. He was already beginning the vigil that ended Monday afternoon."[10]

Barbados native Samuel A. Parfitt, a supervising member of the stokehole crew, told the Steamboat Inspection Service board of inquiry that one hour after the *Vestris* left New York, water was coming through an ash ejector below the waterline, and several hours later he noticed a distinct list to starboard.[11]

Testimony given by three of the stokers, John Boxill, John Morris, and Gerald Burton, supported Parfitt's account.[12] Commander Simpson, navigating officer on the battleship *Wyoming*, said:

> According to statements made by the rescued stokers, the first leak in the *Vestris* came from a cracked sea valve which went down to the ash hopper in the stokehold of the steamer. According to the testimony of the stokers, this sea valve was cracked before the *Vestris* started her final voyage. [They] said there had been some question before the *Vestris* sailed as to whether she would sail at all.[13]

The problems with the *Vestris*, like Carl Sandburg's famous fog, arrived on little cat feet. But it would not be long before this cat broke into a dead run.

≈ ✿ ≈

Chapter Two

The Sea Comes Calling

I T IS DIFFICULT TO EXAGGERATE THE DEEP RESPECT sailors have for the ocean. They refer fearfully to the bottom of the sea as "Davy Jones' Locker," a place alleged to be the final repose of drowned sailors. It is no surprise that the one thing seamen fear above almost everything else is the sight of water pouring into their ship. And that is just what the crew of the *Vestris* saw on Sunday, November 11, 1928.

Third Officer Herbert Welland discovered the first major leak at midnight Sunday when he saw water coming through the "half door" (also referred to as the "working door" and "coal port") on the starboard side, about six feet above the normal waterline. The leak grew steadily worse.

On steamships of that era, a coal port was used to load coal into coal bunkers. The coal was used to fire boilers that made the steam used to drive the propellers of the ship. Firemen shoveled coal from the bunkers into the boilers through a stokehole. The ashes from the burnt coal were disposed of through ash ejectors.

Twenty-five-year-old Second Officer Leslie Watson, who already had six years' experience as an officer, testified:

> The ship was perfectly upright when she left the pier and during Saturday night. When I took the watch at midnight, the wind had strengthened and the sea was rougher. At two o'clock Sunday morning she was listed slightly on account of the weather. She had a five-degree list, or rather heel, at that time. The weather had been getting worse rapidly, and at two o'clock I reported [these things] to the captain.[1]

Several members of the crew agreed that the *Vestris* was doomed by the water that came into her. The coal port could not withstand the huge waves that broke against the ship's starboard side. They said the rubber gaskets that were meant to make the two swinging doors watertight had long ago rotted away and disintegrated, leaving gaps wide enough to admit tons of water.

The possibility that the *Vestris* was in an unseaworthy condition when she sailed from New York was revealed at the Tuttle hearings on November 17, 1928. Chief Officer Frank W. Johnson said he was responsible for seeing that the coal ports were properly closed and fastened, but he admitted that he failed to inspect them before the ship sailed. Chief Engineer James A. Adams testified that the worst leak "may have been" in the starboard coal port. Crew members blamed this leak as "chiefly responsible" for the loss of the *Vestris*.

Johnson had been promoted to chief officer the day before the ship sailed after previously serving as first officer of the *Vestris*. The chief officer was responsible for the seaworthiness of the vessel, including closing and bolting the coal ports before the ship sailed. He told Mr. Tuttle that he did not do this himself but had assigned it to a dependable carpenter and carpenter's mate, who were supposed to have bolted the doors and caulked the cracks. Although the coal ports closed from outside the ship, he admitted that he had not inspected them from the outside since serving on the *Vestris* and did not know whether they had places for gaskets to make them watertight. He told Attorney Tuttle that he did not believe any water came in through the coal ports "because they were bolted and fixed up good." When Tuttle asked how he knew this without looking at them himself, he replied, "You have to depend on someone."

Four o'clock Sunday morning

§ Barbados seaman Fred Gill was called about 4:00 A.M. Sunday to wash down the decks and found water pouring in through the port side coal port doors. He reported this to his superior, who notified the ship's officers. He did not know how long the water had been coming in. It was flowing through an open hatch and into the coal bunkers. He said that eventually the hatch was covered, but water kept streaming into a passageway through the crew's quarters, and from there it flowed into the coal bunkers.

Gill said that repairing the coal port proved impossible. He added that the doors were not opened while the ship was in New York, and he did not know if they had ever been tested. As the water poured through the coal port, the ship began to list, tilting to a greater angle hour by hour.

Some details of the events on that fateful day were supplied by Evans Hampden, a fireman on the *Vestris*:

> When I came on duty in the stokehold at 4:00 A.M. on Sunday, I heard one of the coal trimmers shout that the coal bunkers were full of water. He called it to the attention of the chief engineer, and pretty soon a boatswain's mate was sent down to find out what was causing it.
>
> He found that the coal port on the starboard side—what we call the half door—had been left open about six inches. The door should have been closed, locked, and sealed up tight before we left Hoboken Saturday, but it wasn't. That's what caused all the trouble. He tried to close the door, but he couldn't budge it

because the bolts were all rusty. Two men with sledgehammers tried to close it, but they couldn't do it either. Then he just moved some empty boxes against [the door], which seemed like a lot of foolishness to me.

The water kept pouring in, and pretty soon it covered the floor of the boiler room. The chief engineer had the pumps manned and even helped pump himself, but the water kept coming in faster than we could pump it out. By the time I went off duty at 8:00 A.M. Sunday, it was up to my ankles.[2]

Second officer Watson said that at 4 A.M. he was relieved by the first officer. He then inspected all of the decks and found no damage. He went to his quarters, came out again at eight o'clock, and did rounds with the chief officer during which they found no damage.

Nine o'clock Sunday morning

At the Tuttle hearings on November 17th, Chief Officer Johnson said they could not examine the coal bunker to see if it was shipshape before the *Vestris* sailed because it was full of coal. He admitted hearing water running underneath the coal about 9:00 A.M. Sunday but denied statements by crew members that the coal bunker had begun to fill with water long before 7:00 P.M. Sunday. The chief officer said minor leaks started in different parts of the ship on Sunday.

Chief Engineer James Adams said the first leak was found about 9:00 A.M. Sunday in the starboard ash ejector. It was plugged by noon after letting twenty tons of water into the stokehold bilge, nearly filling it. At 10:00 A.M., the second leak was discovered in a lavatory, which was caused by the carrying away of a scupper plate on the starboard side. This was also plugged by noon after letting fifteen or twenty tons of water into the engine room bilge, which it almost filled.

About 10:00 A.M., Adams found a leak in the port half door of the coal bunker: "It was leaking so badly the coal was saturated." He believed that the storm had sprung the packing around the door.

At the Board of Trade inquiry, Second Officer Leslie Watson testified:

> [I] remained on watch until 4 P.M.; the vessel was then laboring quite a bit. When [I] went off watch, I met the chief officer, [who] told me that water was coming in from the starboard port. After going round the deck, we went down to the cross alleyway. . . . When [we] reached the cross alleyway [we] found that a little water was coming through the starboard door, but it was not bad at the time. The scupper alongside was taking away all the water that was coming in.[3]

Four o'clock Sunday afternoon

By Sunday afternoon, the situation had changed drastically. At four o'clock, Chief Engineer Adams realized there had to be a more serious leak on the starboard side of the coal bunker because of the way the ship was listing. He said the leak was, "possibly [from] one of the coaling doors, or a scupper plate may have carried away in the bunkers, because the ship was straining and pitching heavily."

Adams also testified that there were three other leaks in the interior of the ship, in the bulkhead between the coal bunker and the engine room. Two of these leaks were in the coal bunker, one in a port half door, the other in a starboard half door. They could not be reached because the bunker was full of coal. Tons of water rushed through the leaking coal bunker and through a stokehole into the engine room all day and night Sunday, but much of it was pumped out. In total, four leaks had been discovered on Sunday. Adams believed that these leaks were the cause of the final sinking of the ship. He thought the worst leak may have been in the starboard coal port, which the crew also held to be mainly responsible for the flow of water. Thomas R. Edwards of Liverpool, who was a bedroom steward in first class, stated that the top half of the half door was "always between four and five inches open."

Chief Officer Johnson said the water got into the coal bunker from a half door in the side of the ship aft of the starboard coal port, instead of from the port itself. The half door opened inward instead of outward, like the coal port. He testified that he inspected the half door at 9:00 P.M. Saturday and found it properly battened down. On Sunday, however, he found it was leaking badly. He supervised unsuccessful efforts to tighten the half door Sunday night.[4]

Seaman Evans Hampden talked about what he experienced Sunday afternoon:

> When I came back on at four that afternoon, it was much worse and only a little while after that the water put out the fire beneath the starboard boiler, leaving us only the main and port boilers. It got so high that every time the ship lurched the rush of water knocked men down.
>
> Finally it got so bad that the chief engineer had life-lines rigged to the main and port boilers and firemen were tied to them to make a bucket brigade for passing the sacks of coal along to the boilers. The list was so bad that we could hardly have stood up even if it hadn't been for the rushing water.
>
> They sent down room stewards to bail when they saw that the pumps were choking and couldn't handle the water, and after a while they had firemen bailing, too, but it was like trying to bail the Atlantic Ocean. You just couldn't do it.[5]

About 7:30 P.M. Sunday night: giant waves hit the *Vestris*

Able seaman Alexander Crick, who had been at sea for thirty-five years, described one of the waves that struck the ship as being "nearly as high as this building."[†] The spray from this wave, he said, reached the crow's nest.[6]

[†] The Great Hall of London's Royal Society of Engineers building, where he was testifying at the time—a four-storey building roughly fifty feet tall.

Chief Officer Johnson denied testimony by one of the radio operators that the *Vestris* had a ten-degree list Saturday night. He said there was a slight list caused by the gale from the port side beginning Sunday morning, but that the only reason the *Vestris* hove to Sunday noon was that the weather was so bad she could not steer. The officers first discovered something was wrong at 7:30 P.M. Sunday, he added, when the ship took a heavy lurch as cargo shifted in the No. 1 hold, three automobiles weighing ten tons and some other cargo breaking through a two- and one-half-inch bulkhead and moving fifteen feet to starboard.

According to Third Officer Herbert Welland, there was a "moderate gale" at 8:00 A.M. Sunday, and at that time the ship rolled eight or nine degrees to starboard but came back to an even keel. It was not until the cargo shifted Sunday night that the ship did not come back to an even keel and the list became permanent. But others said there was a notable list even before the cargo shifted.

Officer Welland confirmed that cargo shifted at 7:30 P.M. Sunday night consisting of automobiles and packing cases weighing about fifty tons in the No. 1 hatch. He said the ship had been struck by a tremendous sea, causing the cargo to break through the fore and aft bulkhead and shift fifteen feet to starboard.

Second Officer Watson described the events of Sunday afternoon:

I went back on watch at noon, when the ship was hove [to]. During the afternoon we ran the starboard engine on slow and half speed. At four o'clock I was relieved, and inspected the decks again with the chief officer. We went to the cross alleyway, where there was only a slight amount of water coming in the half door at that time, and the small bunker hatch was securely battened down. I reported to the captain that I had found no damage to the decks. The weather was still worse—it was a gale then—and I was with the chief officer in various parts of the ship until seven o'clock [Sunday evening].

At 7:30 P.M. we were at dinner, myself, the chief officer, and the first officer, when the ship took a heavy lurch. I immediately went on deck to see what damage had been done and found the rail on the starboard deck had been carried away. I received the report that the cargo had broken through the bulkhead [at] No. 1 hatch into the crew's quarters, and I went to inspect that. I found it could do no further damage; it had shifted as far as it could, and nothing else could move. Then there was a port broken in the firemen's forecastle. I had the carpenter fix that.

The ship had been rolling evenly before that lurch, but after it she never came back. She rolled then between five degrees to starboard and fifteen degrees. From eight o'clock Sunday night on I carried orders continually from the captain to the chief engineer. They were orders concerning the ballast tanks, and questions as to how things were in the engine room. The answers which I brought back were all reassuring answers.[7]

Clearly Captain Carey was being misled.

MONDAY AT 4:00 A.M., CAPTAIN CAREY told Watson to muster all hands and form a bucket brigade in order to supplement the pumps. The survival of the *Vestris* was balanced now on a razor-thin line.

~ ❀ ~

Chapter Three

The Passengers Try to Cope

DOCTOR August Groman of Odebolt, Iowa, aged seventy-two in 1928, was the oldest survivor of the sinking of the *Vestris*. At the time, he was a veteran of several sea voyages; hence, when interviewed, he was able to give a very informed account:

§ The ship seemed to be in good shape Saturday. I had no reason to believe there was anything wrong with her. That night, however, a terrible storm broke over the ocean. It became more terrific each hour. The boat heaved in and out of the monster waves. Water began to pour into her body. It was the first time in all of my days on the ocean that I became seasick while in bed. I have been seasick during the daytime, but never in my cabin at night. So I claim it was a very bad storm.

This interview with Dr. Groman, made a mere ten days after the *Vestris* set sail from Hoboken, continued with his remarks about the second day of the voyage, Sunday, November 11.

Very few people got up the following morning. The reason was that many of them had also become sick during the night. I noticed, however, that the ship was listing Sunday morning. The list became gradually worse during the day, and by night the ship must have been listing some twenty degrees.

I talked to a friend of mine on board ship about it. Both of us came to the conclusion that something serious must have happened. We found that hardly any food was available during the day, with the exception of some biscuits. The reason for this, I found out later, was that the ship was listing so badly it was almost impossible to cook anything. For instance, a pot of coffee would fall over if placed on a stove. The same difficulty was experienced in trying to serve food.

Of course, this condition was not felt so badly by the passengers as it might have been, because many of them were too sick to eat anything.

When I retired to my cabin Sunday night, the *Vestris* appeared to be in a serious condition. There was a good bit of water flowing through her, but nothing of her actual condition was reported by her officers [or] members of the crew.[1]

This last remark by Dr. Groman starts a theme that will run through the rest of the reports by survivors of the *Vestris*: the apparent indifference of the captain and crew for the well-being and plight of their passengers. A *Time* magazine article captured this situation in the following sentence: "Hysterical survivors filled the press with stories of leaking lifeboats, faulty tackle, indifference of officers, mutinous and incompetent crew."[2]

A *Baltimore News* story dated November 15, 1928—evidently an evening newspaper, as it contained information from the federal hearing which began that same day—supplies a great deal of the testimony from passenger Fred Puppe, whose wife and seven-month-old child were lost in the sinking of the *Vestris*.

§Puppe, the pain of his loss showing plainly in his face, testified in a breaking voice. He said that on Sunday he had risen at 6:00 A.M., the time when the baby was usually fed. The ship, however, was at such an angle that he found it hard to move around in the stateroom. His wife was also unable to move about comfortably in order to prepare the baby's food, so Puppe sent for a steward, intending to order some oatmeal.

No steward came, so Puppe said he started to prepare the cereal himself on a hot plate. It spilled a couple of times, he said, showing United States Commissioner O'Neill how he had tried to keep the pan on the cooker. Puppe said his wife was feeling sick, so he went to breakfast alone.

> I thought that the ship's incline was always the same way and that something must be decidedly wrong. I've travelled a great deal, have been across the ocean twelve times, and up and down the American coast itself a great many times, and I know when a ship stands on one side and never turns over to the other side, something is wrong.

> I met a steward and told him something was decidedly wrong. The steward said to me: "You don't know anything about it. The cargo has shifted. The crew is working on it now. Everything will be straightened out in about an hour."

At noontime lunch on Sunday, Puppe said the ship was listing so badly that the passengers in the dining room had to hold down their plates on the table with one hand while they ate with the other.

"On Sunday," he said, "my long experience showed me that the list had steadily increased since morning." He added that during the afternoon he was in his cabin most of the time caring for his sick baby because his wife was so ill. In mid-afternoon, he testified, "I went on deck and at that time noticed that the list had increased even more."

That night his wife became so sick he called the ship's doctor. He asked the doctor if he couldn't get some food from the kitchen for the child. The doctor answered that it was impossible to cook in the kitchen any longer.[3]

The passengers may not have known what caused the sudden shift in the ship's attitude when the temporary wooden bulkhead gave way and allowed some of the cargo to crash through to the starboard side of the vessel, but they certainly were aware that something had happened, as witness this vivid narrative by William P. Adams—the traveling companion of Dr. Groman—although the timing given by him is suspect:

> Sunday morning about ten I was reading a book on the port or "high" side of the ship, in the ladies salon, when the vessel made a hard lurch and everyone and everything on that side, including the chairs, tables, rugs, etc., slid down to the low side of the room. At that time a *very* hard storm and *high* sea was running. It seems that Mr. Jackson [Dr. Ernest A. Jackson, a missionary] had been sitting near me on the high side of the room & after we crawled out of the jam of furniture etc. we each got a big over-stuffed armchair and put it on the high side of the debris. Knowing we couldn't slide any farther, [we] entered into a conversation. I told him of my expected fishing trip to some islands off the Chile coast near Valparaiso, and he told me of his life and work in the Baptist missions in Brazil.[4]

Dr. Jackson and his wife, Jannette, as well as their son Cary did not survive the wreck. Another son, Judson G. Jackson, who was not a passenger, sent out letters of inquiry asking for information from survivors of the shipwreck. The quote above was in one of those letters Judson Jackson received in reply.

The shift of the cargo at 7:30 P.M. does not jibe with the version by passenger Adams, who states that the sudden lurch occurred at about ten o'clock Sunday morning. Since Mr. Adams described a "very hard storm and high sea" at the time, which actually happened

Sunday evening, it is likely that he was simply mistaken about the timing of the lurch.

Captain Frederik Sorensen, another passenger, gave this report:

> Sunday night we all had to eat in the first-class dining room because the fire in the second galley was out. Some of the rooms were swamped. There [were] not many in the dining room. The ship was rolling heavily and nearly everybody was sick. Some of the waiters got hurt in going around, and one got his finger cut off. I had a good appetite. I was hungry. But finally I had to give up and just asked for a piece of meat and some vegetables on one plate.
>
> Just then the *Vestris* took one heavy roll and everything went to the corner—tables, chairs, china, and food. I think some of the passengers got hurt, and some of them gave up then. There were some women in the dining room. They got out when that happened. A few of them started to cry but no one screamed or carried on much.[5]

Again, the timing is in the evening, not in the morning. Carlos Quiros, who was quoted earlier, had this to say about the matter:

> At seven o'clock, when it was time for dinner, I saw that there were only three tables in the dining room occupied and I decided to go to my stateroom. I rang for the steward and ordered consommé, eggs and fruit. He brought the order and I began to eat. Just then the *Vestris* gave a violent roll. The soup and eggs and fruit and I fell onto the floor. Most of the furniture and my baggage came tumbling across the room. It was seven thirty or seven thirty-five o'clock Sunday evening.[6]

Quiros seems to have the timing down rather accurately. Another passenger, Dr. Ernst Lehner, who was thirty-four and came from Basel, Switzerland, had the timing of this event a bit later than the others but paints a credible picture of the scene:

The storm gradually increased, and about eight o'clock it was at its very worst. We had dinner then—a very hurried meal, just a few scraps of food. The first really big wave that I noticed hit the ship just while we were at dinner, and one steward got knocked over so badly that his face and hands were severely cut. I went out after that, and just as I reached the promenade deck it was struck by a wave that flooded it and the Winter Garden inside, knocking over all the chairs and a few people at the same time. There were very few passengers about then—I should say, twenty in all—and they kept under cover. The third-class passengers apparently were locked in, because we did not see a sign of them on Sunday while the storm was going on.[7]

Fred Puppe, in his testimony before the federal board of inquiry, described the weather conditions preceding the sinking, in which he stated that the sea on Sunday was not as heavy as he had seen it on other trips. "I felt no particularly strong wind—certainly no wind strong enough to heel us over, as the stewards said. The waves were decidedly smaller than I had seen in previous storms."

Passenger Frederik Sorensen, who was also a sea captain, told of problems he witnessed that developed by Sunday afternoon:

The bad weather really started Saturday night, and when the ship listed the next day, it was at an angle of about 40 degrees. We couldn't get any information as to the trouble from the captain or crew, and when the engines had stopped and [we] were lying helpless Sunday, the captain said there was no trouble and we were proceeding at once, which was not so.[8]

It turned out that Sorensen's estimate of the ship's list early Sunday was wide of the mark, as the list that morning was only about four or five degrees; he may have meant Monday morning.

Jorge doValle, a third-class passenger who was formerly attached to the Brazilian consulate in New York, was of the opinion that the ship never should have put to sea, saying, "The ship was in bad condition when she sailed. When she started to list, water came in through leaky portholes and everywhere. The rubber fittings were missing. A lot of water got into the coal bunkers. The pumps were working the whole time, but faster than they could pump, the water came in."[9]

Businessman Hermann Rueckert, 26, of Leipzig, Germany, said that after the list grew worse Sunday afternoon, he had to change cabins three times because of water coming into the rooms.[10]

Orrin S. Stevens, of Buenos Aires, dramatically told his version of the events of Sunday evening:

> We met rough weather all day Sunday and about seven Sunday evening a huge wave hit us and started a leak. The portholes were already leaking and shipping water.[11]

Edward J. Walsh of 137 Eighty-eighth Street, Brooklyn, an auditor for a construction business in the same city, spoke of the growing danger to the ship. Early Sunday morning the *Vestris* ran into rough weather, and about 11:00 A.M. it hove to, Mr. Walsh said. He understood that the coal had shifted and water was leaking through the portholes.

He described his predicament in these words:

> About three or four Sunday afternoon, I realized that we were in grave danger. The ship's list kept getting worse and worse, and about 6:00 P.M. it was so bad that the tables and chairs, which were fastened down, ripped loose.
>
> I sat in my stateroom all night Sunday just waiting for it to happen. The next morning the list was so bad that you could not walk on the decks. You had to make your way clinging from one thing to another.[12]

A graphic tale of the events of Sunday

The following review of the leaking *Vestris* was written by survivor Dr. Ernst Lehner, who was en route for Trinidad, where he was the chief geologist for Trinidad Leaseholds Ltd., an oil company.

When the ship listed to starboard early Sunday afternoon, I did not become afraid because I thought it was due to the wind, which was hitting us broadside. It seemed to me the boat was going just like a sailing boat, and although I am a bad sailor I did not get seasick then at all.

Before lunch that day the ship stopped, and it did not seem to get started again until about three o'clock in the afternoon. Johnston[†] came up to me and said: "Look here, Lehner, I am glad to see this get going again. Whenever these people stop a boat there is something the matter."

. . .

I was not frightened for the simple reason I knew nothing about it. I always had the impression that the ship was taking the storm very well. There was only one woman, a Mrs. Stevens who was later lost, who seemed much upset to me. She was dreadfully frightened and seemed to have a presentiment of what was coming. Her husband and I had a deuce of a time comforting her. There were also some German fellows huddled up in the corner of the Winter Garden, oblivious to what was going on.

We never saw anything of an officer during the whole of Sunday. They must have been on the job.

As I did not sense any danger I made no inquiries. At about nine o'clock I went down to bed and slept very comfortably. Only once in the night did I wake up, and that was to hear a fellow outside say: "Well, she isn't any worse than she was four hours ago." And then I dozed off again.[13]

[†] Forty-four-year-old Herbert C. W. Johnston, who was traveling with Lehner, was a general manager for Trinidad Leaseholds Ltd.

Once again, we hear that passengers were not aware of any danger on Sunday, despite the rather spectacular chaos in the dining salon when the big waves hit around seven-thirty that night.

Another passenger, thirty-five-year-old Paul A. Dana of Boise, Idaho, who worked for the RCA Corporation in Buenos Aires, was more keenly aware of the danger presented:

> Our first night out, Saturday night, the *Vestris* began to hit rough weather. As the night progressed the storm got worse until, before the night was over, we were in the worst storm I ever saw on the sea.
>
> It was late that night, perhaps a little after midnight, that a thing happened which I believe started the trouble that ended in the tragic sinking of the *Vestris*.
>
> Two big waves struck her, almost simultaneously, bow to stern. The ship quivered from end to end. You could almost feel her wrenching. The next day she developed a leak. One of her plates must have been wrenched loose.
>
> It was rough Saturday night, and Sunday was rougher. Only four of the passengers besides myself were down for breakfast. The minute I stepped out of my cabin that morning I began to feel uneasy. The *Vestris* was listing. I had been on steamers before that listed, but I had never seen quite such a list before. It looked bad.
>
> In the dining salon that morning I ran into Captain Frederik Sorensen, and we started to talk things over. He did not like that list either. Inasmuch as he was a sailor I decided he must know what he was talking about and that he was a good man to stay with. We spent most of the day together.
>
> Whether the other passengers were uneasy I don't know. Most of them were violently seasick and spent the whole day in their bunks. As I said, there were only five down for breakfast. No more showed up for lunch or dinner.

By the middle of the afternoon, while the storm still continued with no apparent letup, the list had become so pronounced
that all the furniture that wasn't fastened down in the dining
salon and in the smoking room had slid over to the starboard
side, where it was crashing around as the ship rolled.

After dinner, Sorensen and I went into the smoking room and
started for the bar. We managed to find two chairs wedged in on the
starboard side of the smoking room, from which we could reach
around into the bar. I had a couple of whisky and sodas, felt a little
better, and went to bed about ten o'clock. After wedging my bed into
a corner of the cabin so it couldn't slide around, I got to sleep.[14]

Though Dana gave a good graphic description of the event, he—like
passenger William Adams—mistakenly placed it as early Sunday
morning, whereas it was actually in the evening. The giant wave—or
waves—did considerable damage to the ship and her workings.
Apparently no plates were "wrenched loose" by those enormous
waves, but the waves carried a large amount of water on board,
which caused the *Vestris* to heel more than usual and allowed water
to come in through damaged ports and the half door.

Passenger Carlos Quiros contributed this written statement:

Then about nine-thirty [Sunday night] the sea calmed down.
I had gone back to my stateroom. I was quite hungry now and I
called the steward again. I asked him to bring me a half a bottle
of champagne and some fruit. He came back quickly to say that
everything was upside down in the pantry and in the kitchen and
the bar, but he brought me some grapes and a drink of gin and
water. I asked him if there was any danger. He assured me there
was not. He said the crew was pumping out the ship and that the
sea had calmed down. I knew that to be so and it was fortunate
for had the storm increased the *Vestris* could not have lasted two
hours. Reassured by the steward, I decided to go to bed. I changed
beds to be more comfortable, but even so I did not sleep.[15]

Two other survivors, Thomas E. Mack of Tekla, Wyoming, and his friend Ovelton L. Maxey of Richmond, Virginia, both employees of an oil company, gave this version of conditions on Sunday:

> Twenty-four hours before the ship sank, everyone on board knew that something was seriously wrong but nothing was done to prepare for possible disaster, and the officers refused to discuss the situation with the passengers as if they were not concerned in the fate of the steamer, the two men said. Mr. Mack bitterly remarked that if SOS calls had been sent out in time, nearby ships could have reached the *Vestris* before she sank and would have saved all on board and even their belongings "with ease."
>
> The two men said that shifting of cargo was explained by officers as the cause of the ship's list when that condition was observed by the passengers on Sunday afternoon, but only when the same question was persistently put to them by everyone on the ship. They added that in their opinion the cause of the ship-wreck was due to the intake of water following the smashing of several of the ship's plates by the storm.[16]

Dr. Groman contributed this description of the events of that stormy Sunday night, including contact with his friend, sixty-five-year-old Chicago millionaire William Phipps Adams:

> Late that night, while I was sound asleep, her engines stopped. I had been so accustomed to the rumbling of the motors what when they were shut off the quiet immediately woke me. I knew then that something still more serious had developed, but I remained in bed.
>
> I did not look at my watch, but it must have been about four o'clock in the morning when my friend came to my cabin and knocked on the door. I unlocked the door and he entered. He had a worried face. He was fully dressed. I asked him what was the matter, and he told me that he thought I had better get dressed because he believed the ship was sinking.[17]

Indeed, the hours left to the *Vestris* were now numbered in single digits. It was time for all on board to begin thinking about what they would do if the ship were to go to the bottom of the sea.

~ ❀ ~

Chapter Four

Pandemonium Breaks Loose

WHEN MONDAY MORNING BROKE OVER THE VESTRIS, the ship was listing about twenty degrees to starboard. As the passengers awoke, they quickly realized that something was seriously amiss. After Dr. Groman's friend, William Adams, told him that he believed the Vestris was sinking about four o'clock Monday morning, Groman said, "I told him I would put on my clothes, and he left for the deck. I found it hard to get up. . . ."[1]

Judge Goddard wrote in a 1932 opinion that on Monday morning the wind had gone down and the weather was fair, although there was a heavy swell. Officers from other ships in the vicinity reported winds that reached Beaufort 12, which is wind of hurricane force. The storm was not, however, classified as a tropical storm. He wrote that the storm that the Vestris encountered was no more severe than is reasonably to be expected at that time of the year. Other ships in the vicinity of the Vestris, including some rather small craft, did not have any trouble weathering the storm. His conclusion was that overloading and poor condition were fatal to the Vestris. (See Appendix A.)

Dr. Groman, whose narrative interview contributes so much to this book, gives a vivid story of the events of Monday morning:

I rang for a steward. In a little while he appeared at my cabin. I asked him to get me some grapefruit, thinking that that was about all my stomach could stand. He looked at me in dismay.

"I'm afraid, sir, I can't get you any grapefruit," he said. "Nearly all of the food has been ruined by the storm."

"Then get me some other kind of fruit. I don't care what it is," I told him.

"I'll do the best I can sir," he replied, and left.

In the meantime I proceeded to dress. There was water on the floor of my cabin. My clothes had been hanging from a hook, however, so they were dry. But my shoes had been on the floor. They were floating in water. I put them on anyway.

In about a half hour the steward returned to the cabin. He brought a handful of grapes he had found in some out-of-the-way corner. Noticing the anxious expression on his face, I asked him what the trouble was on board ship. Tears came to his eyes. He actually cried. "I'm sorry, sir, but I can say nothing," he stated in his English accent. "Orders are orders. I can't talk. God, I'd like to tell you. But I think you know what I'm trying to say." With this, he left once more, and I was not to see him again.

Fully dressed now, I left my cabin and started to walk towards the stairs at the end of the hallway. I could do so only with great effort. I was very weak. Upon arriving at the stairs I reached for the railing, but my arm slipped because of its lack of strength and I fell headlong to the bottom of the stairway into a pool of water several feet deep. I was badly bruised on the side of my head in the fall and was somewhat dazed.

I finally got out on the deck. Most of the passengers were clinging to the railings on the high side of the *Vestris*. Here I also found my friend. We felt then that the ship would undoubtedly go down.[2]

There seems to have been more to Dr. Groman's fall and recovery. In testimony before the British Board of Trade, Captain Frederik Sorensen told this story about the events of Monday morning:

> After I found that I could not get breakfast, somebody finally got a bunch of bananas from the icebox. I carried them up and started distributing [them] round to the passengers, finally getting to the smoke room, and gave out the last of them there. I went down again, and on my way down I noticed somebody had gone through the door in the salon—lost their foothold and crashed out to the lee rail, which was under water.
>
> I got down and helped them up; it was an elderly man, a passenger, and a couple of waiters. I think the passenger had crashed through the port and the waiters got out to help him. They got caught by the sea and all were struggling and scrambling out there. One of the waiters had a broken nose; the other one, I don't know whether [he] was the one that cut his wrist. I got full of blood when I grabbed him, but we got them in. I was in a pretty weakened condition, having just come out of hospital. Still I managed to get around pretty good; I suppose the situation sort of braced me up a bit.[3]

Fred Puppe, the forty-five-year-old German who was traveling with his wife and young son, continued his testimony at the Tuttle hearings concerning the events of Monday morning.

§He said that at 7:00 A.M. Monday he wanted some food for his wife and child, but was told that none was being served.

> I went down to the kitchen myself to see if I could find anything and found a fountain four feet high. The portholes were closed, but every time a wave hit them the fountain sprang out. I realized the same thing must be happening lower down. But when I asked for something to eat, I was told "go get it yourself." I finally got two bananas.[4]

Later, he testified, his wife wanted some water. When Puppe asked
for it he was again told "get it yourself." He could not find any.

By eight o'clock Monday morning, I was absolutely sure that
an SOS had gone out hours before. Anyone with the lives of so
many persons on their hands should have called for help long
before. I never thought for a moment that there hadn't been a
distress signal.

Puppe's voice faltered here. He took off his glasses and polished them
with a handkerchief before going on with his story:

When I saw there was no hope anymore, I took my wife and
baby to the smoke room and later to the deck. This was at 9:00
A.M. We waited on deck, looking for the steamers we were
absolutely sure must have been called to our help.

Other passengers were asking if an SOS had been sent, but I
was absolutely sure it must have been.

Suddenly, though we heard no orders and although no
officers were in sight, the crew began to take down the lifeboats.
You could see that none of them had ever even tried to lower a
lifeboat before.

There was an absolute lack of knowledge [of] seamanship.
They ran from one boat to another, taking things from one and
putting them into another. I didn't realize what this meant, but
later I discovered [what was going on].

The witness testified that the list of the *Vestris* Sunday morning was
ten degrees or less. During the afternoon it was eighteen to twenty
degrees, he estimated, and Monday morning about thirty degrees.

"Monday morning between six and seven o'clock," he said, "I asked
a steward why the engines had stopped. That steward had the nerve to
tell me that it was because we were using all our power for pumping."

Attorney Tuttle then asked him, "Did any of the officers or
members of the crew at any time give you an explanation of the
cause of the list other than that the wind was pressing you over?"

"No," answered Puppe. "All the explanation I got was what the steward said—that the wind was pressing us over on one side."

On another subject, Tuttle asked, "Did you ever see the captain or any of the officers give any command to any one of the crew that was not executed?"

"No," answered Puppe. "I never once saw any officer give any command. The only thing I did see was the First Steward order some of the stewards to go after food for us. They refused flatly. One of them said, 'I wouldn't go back there for a thousand dollars.'"[4]

Mr. Puppe may have over-estimated the degree of list at the times given, but his testimony is very descriptive. *Time* magazine, however, gave only a sketchy account of Monday morning's situation:

> By dawn Monday the gale was but a whisper, the sun burst through a sky of scudding rain clouds. But passengers on the starboard side, looking out of their windows, could not see the horizon. "It was like looking down a deep well." The deck tilted like a barn roof.[5]

Judge Goddard in his 1932 opinion wrote that between three and four o'clock Monday morning, the cover on the starboard hatch in the cross alleyway leading to the [aft] cross bunker was blown up by the pressure of water in the bunkers. By 4:00 A.M. Monday, the water had risen so high that the starboard boiler was closed down, and although her pumps—with the exception of the circulating pump of the engine—were in operation much of the time, the water was gaining and a bucket gang was organized to bail water from the cross alleyway. This did not work and was discontinued at eight in the morning. At 8:37 A.M., the alarm signal CQ was sent. This meant that all commercial radio traffic was to cease pending the receipt of an SOS. At 9:56 A.M., the SOS signal was sent. This was answered by fifty-eight ships and a number of shore stations. Shortly before 1:00 P.M., the water-tight bulkhead separating the engine room from the stokehold burst, and soon after this the engineers left the engine room and went up on deck because the engine room, stokehold, and bunkers were flooded. There was nothing more they could do.[6]

In an interview, Samuel Parfitt, who was the lead fireman on the *Vestris*, gave this account to a reporter:

> I was on the promenade deck and I stood as close to Captain Carey as I am to you. He was bareheaded. I saw some officers come up to the captain. "What about the crew, Captain?" asked one. In an angry voice [the captain] snapped: "Damn the crew!" Then he just stood there doing nothing. That was about 1:00 P.M. [Monday]. In my opinion the heavy loss of life was due to the negligence of the captain. If he had given the proper orders, many more would have been saved.[7]

AT THIS POINT there was absolutely no doubt that the *Vestris* was done for. She would not last even another two hours.

Chapter Five

A Few Good People

T HE CAPTAIN OF THE *Vestris* WAS A STUBBORN MAN. His vessel had been listing to starboard for more than twenty-four hours, and still he refused to issue an order to abandon the ship. *Time* magazine wrote: "At nine o'clock, Captain Carey, hitherto indifferent to pleas of passengers to 'do something,' ordered women and children on deck."[1] But he still could not bring himself to order the passengers and crew into the lifeboats. Finally, at 11:40 A.M., Carey gave the order to lower the lifeboats.

Time magazine reported that despite the situation most passengers did not panic. They said that some of the crew began looking out for themselves, ignoring both officers and passengers. When an officer pointed his pistol at one of the rampaging crewmen, the seaman grabbed the weapon and tossed it into the sea. Captain Carey gave the then-outdated order: "Women and children first!"

Time quoted Anthony Lewkowicz, who designed the lifeboat davits and falls on the Vestris, as telling a group of reporters that his lifeboats were "unsinkable" and the tackle "foolproof."

Lewkowicz stated: "With my davits a boat with a full load can be launched safely by one man . . . in spite of a thirty-two degree list. . . . The average time is fifteen seconds." Nevertheless, some of the lifeboats did sink, and davits malfunctioned. A few lifeboats took two hours to launch and others never got free of the ship.

Most of the women and children were sent to lifeboats on the port side of the *Vestris*. There, the clapboard construction of the lifeboats caused them to catch on the hull plates of the *Vestris*, and they proved difficult to launch. One of the falls on No. 4 broke, dumping its passengers into the raging sea. A davit broke, dropping its heavy arm into lifeboat No. 2 and killing many of its occupants instantly. It broke through the bottom of the boat, which then immediately sank.

For two hours the crew struggled mightily to free No. 6 from its falls, without success. *Time* wrote: "When the *Vestris* nosed under, No. 6, still fast, was dragged down with her; and a third boatload of women and children was strewn upon the sea.

"Women and children were first—to drown."[2]

Before the order to launch the lifeboats was given, there were some acts of quiet heroism. Some of these were attributed to the missionary Ernest Jackson as learned in two letters sent in reply to the circular mailed by his son Judson Jackson requesting information about the fate of his parents and brother, referred to previously. (Images of all of these letters are reproduced in Appendix C.) Captain Frederik Sorensen wrote this:

> Mr. and Mrs. Jackson's picture has remained in my memory. I feel quite sure on account of the following: Shortly before abandoning the ship a number of passengers were gathered in the smoking room. Mr. Jackson was sitting with his wife. Suddenly Mr. Jackson spoke up in a loud clear voice something like this:

"We must now pray and trust in God to help us out." That was the
last time I remember seeing them.[3]

Another letter written to Jackson was from Dr. Ernst F. Lehner and
reads in part:

> I can vividly recall the picture of your parents and Cary sitting
> in the smoking room on Monday morning. They were very quiet
> and composed. We conversed on some trifling matters in the early
> morning hours. When it became certain that we would have to
> take to the boats, your mother sent Cary down to the cabin to fetch
> some of her things, which he brought up in a small handbag. After
> that both your father and Cary did splendid work in collecting life
> belts from the cabins, which they distributed among those people
> who were too frightened to get them themselves. When everybody
> in the smoke room had their life belts, your father started praying
> in a low but firm voice, comforting so the little group, which by
> now was augmented by some colored women and children from
> the second and third class.[4]

The Tuttle hearings before Commissioner O'Neill

[§]District Attorney Charles Tuttle began his investigation November
15, 1928, into the responsibility for the sinking on Monday of the
Lamport & Holt steamer *Vestris* before US Commissioner Francis
O'Neill. This investigation is herein referred to as the "Tuttle hearings." Not interested in settling questions of jurisdiction, DA Tuttle
wanted to get the stories of survivors on the court record, asserting
that it was most important to "perpetuate their testimony."

Six witnesses who were passengers on the *Vestris* testified. All
agreed that during the crucial time before the ship sank, there was
no organization or discipline among the officers and crew. They said
that no orders were given and that it was the passengers rather than
the crew who seemed to be calm and collected.

Fred Puppe was the first to testify, stating that he was an electrical engineer and had been traveling to South America with his wife and seven-month-old baby to a position with an electric utility company there. In his testimony to Attorney Tuttle, Puppe said in part:

> At dinner Saturday night the ship was sailing smoothly and evenly, and passengers remarked to each other that if the whole voyage continued thus it would be an excellent passage.
>
> . . . I woke up in the middle of the night and noticed that the engines had stopped. I thought it was about two o'clock. My wife woke up frightened and I reassured her. I told her that I believed the captain was a capable man and that if anything was really wrong a dozen ships would be standing by us long before they were needed.
>
> Some of the lifeboats never got away at all, due to the lack of facilities for lowering them. I saw two boats still hanging on the ship when she went over.

"What side were they on?" asked Attorney Tuttle.

> The [port] side that was out of the water. They were resting on the ship's hull. I could not identify them, but I could see people in them.[5]

Another witness was John Santana, a Brazilian elevator operator who was returning to his native land as a third-class passenger on the *Vestris*. He said there was a storm Saturday night, and at six o'clock Sunday morning he noticed that the ship was listing. The list had grown so pronounced by evening, he said, that the dishes slid off the tables in the dining room. He was rudely awakened that night by a crash that threw his luggage across the cabin. He thought it might have been from a shifting of the cargo.

Going on deck at six o'clock Monday morning, he found the list had grown even more. He saw the crew throwing cargo overboard, so he asked a steward what was wrong. He was assured that everything was all right. But three hours later the women and children in the third class rooms were ordered to the first class deck, and he realized that the ship had a serious problem.

Santana said that about noon, they started putting the women and children into lifeboats. Half a dozen times he tried to get into a lifeboat, but none of them would let him in. Once, he saw two boats full of women and children that had not got down to the water, but were stuck on the side of the ship, hanging by their falls about a meter above the waves.

While he was running along the deck, without warning the ship lurched and rolled heavily over on her side. He barely had time to leap over the rail and then found himself running on the side of the hull, which was almost level for a few seconds. As he ran for his life, he saw the two boats of women and children still stuck on the side of the ship. They had been caught up on the side of the hull and were tilted at a crazy angle.[6]

Alfred "Fred" Hanson, an assistant pastry chef, testified at the Tuttle hearings in a voice so weak and low that only the stenographer could hear him. She had to repeat his answers for the court. When Tuttle told him to speak louder, he said his voice had not recovered from the night in the waterlogged No. 1 lifeboat.

Hanson admitted knowing that water was getting in through a port of the ship. Sunday night was very windy, and the ship started rolling. One of the starboard ports leaked, and the water got down into the coal. He stated there was even water in the pantry where he worked, and his supervisor fell and cut his hand trying to get about.[7]

§Monday morning at about three o'clock Second Steward Duncan came down and called everyone and told them to bail out water because the water was all over—even in the dining room and in the room where I slept. We got hold of some water buckets; we were about fifty men bailing water until nine o'clock in the morning when we were told it was no more use.

I went into my room and got my camera and went on deck to take some pictures. There were only about three men on deck. The chief steward [Richard Davies] was on the bridge looking down. I went to a lifeboat in the stern and took some pictures.

Finally, I went into the pantry and had a couple of bananas, and we were seated for about two hours. Then they called the crew on deck to launch the lifeboats. We started to work on boats 4 and 6. When the davits were swung out, the boats were left inside the railing, and it was very hard to get them out. One man [Steward George Hogg] broke both his arms.

We worked from eleven until two o'clock in the afternoon on those two boats. Meantime numbers 8, 10, and 12 boats were sent down. There was no use trying to get those other two boats down, so I went and took some pictures on the deck.

At twenty minutes to two, those boats were still standing there. We were working there with poles to push them off. We pulled them up several times to get them down, but it was no use.

Tuttle broke in, asking, "And you never got them in?"

No. When the ship went down, one went down with it, and the other was left floating upside down. Almost all the people in them were drowned.

Hanson explained that they were unable to lower the two boats, which were loaded with women and children, because the list of the ship caused the boats to contact her side and catch on her plates.

Hanson handed over a group of the pictures he had told about taking, and Attorney Tuttle put them in evidence.[8]

At the British Board of Trade inquiry held in 1929, Hanson produced a complete list of the photographs he took on the *Vestris*. They were made with a Kodak folding camera, which he had bought for $8.50 just the day before the *Vestris* sailed.[9]

A fireman from the *Vestris*, Joseph Boxill, gave this testimony at the Tuttle hearings, which is paraphrased.

Boxill went on duty at midnight Saturday, November 10, which was when he first noted a slight list. Yet it was a minor list, much as he had noticed on previous occasions. He said that the ash ejector valve, the source of a leak testified to earlier, was leaking then. A lot of water was coming in and being carried off into the starboard bilge. He thought the leak was stopped at noon Sunday when the storm was at its worst. The list had become worse, and water made it hopeless to try stoking the starboard boiler.

The starboard boiler was out altogether by midnight Sunday as a result of the water rising on that side. Boxill said the list was "very heavy," the bilges were filled, and water was over the floor plates on the starboard side. Water was coming out of the chute leading from the upper coal bunkers. Firemen stoking the two boilers still working had to be tied together to keep from falling in the water.

When his shift ended at four o'clock Monday morning, Boxill and twelve of the other firemen were ordered to bail water at the leaking starboard half door on deck below the main deck. When Boxill returned to the stokehold sometime after eight o'clock, he found that only one boiler was working.

The firemen all stayed on the job and obeyed orders, Boxill testified. Boxill himself continued working, even though he had gotten nothing to eat since Saturday. He went up on deck at ten o'clock Monday morning and discovered a member of the crew with some soda crackers. He had little time to enjoy the crackers, however, because he was ordered to help throw cargo overboard a few minutes later. When the jettisoning of cargo was abandoned as hopeless, he lay down on a hatch cover to get a little rest. But even though he was sick and exhausted, he was ordered again to go to work in the stokehold when the chief officer ordered all firemen back to work on the boilers.[10]

One survivor, a Mormon missionary named David H. Huish, gave an account to the church elders. Huish was accompanied by Keith W. Burt, who did not survive.

> Monday morning we both went down to get breakfast but all the stewards were working bailing out water, so we went back to our room and stayed until about 11:00 A.M., at which time they sent a boy down to tell all the people to get up on deck and buckle our life belts—saying there was no danger in order to keep the passengers calm. It took us quite a while to get there; sometimes we couldn't climb the slant of the ship, and once I slipped and was thrown against a chair [on] the other side of the room and broke the chair into pieces. We finally reached the deck and had to lean against the wall or hold to the railing of the ship in order to stand up. Even ropes were used to help people from the stairways to the railing. We watched them lower the lifeboats from then until about 2:00 P.M., at which time we went down into one ourselves. They had some difficulty in lowering our boat and were just cutting it loose from the ship, when the ship sank with our lifeboat still tied onto the ship. We all saw this and were forced to take to the water.[11]

Striking in these accounts is the calmness of the passengers in the face of what can only be described as pure pandemonium. The confusion with the lifeboats, the accidents that occurred, and the evidently complete absence of direction or guidance by the ship's officers contrast sharply with the orderly behavior of the passengers.

Another of the letters sent to Judson Jackson in reply to his request for information was that of William Adams. We pick up his narrative following the talk he had with Mr. Jackson in the ladies salon, recounted earlier:

> §After about an hour of this I went to my room and do not remember seeing him [Jackson] again—nor on the promenade

deck the next morning—nor in any of the small boats in the afternoon, but I was at the after end of the deck, which had a list like this [he drew a line on the page here at about a thirty-degree angle from the vertical] so I could do nothing but hold onto the rail or stanchion until I got into a boat with my friend (he is seventy-two and I am sixty-six) and did not walk about nor again go below for that reason.

It may have been that he thought his chances were best on the lee side of the ship—but a raft with fifty people on it and two boats full that were launched last on that side were caught by the superstructure of the ship when it finally rolled over—and I presume many if not all were lost. The next boat to ours on the port side was very leaky; it filled and sank shortly after being launched.

I do not think Dr. Jackson was in the same boat with his wife as this was not allowed and the only order I heard given and the only officer I saw that day was the Captain ordering a man out of the boat that contained the women—he was saved and his wife (on their honeymoon) was lost. That boat was the first loaded, but for some reason was the last to leave the ship; in fact, it never did leave the hull of the ship, for when we were about 250 feet away and were on the top of a wave and the ship—now nearly on her side—was also on top of a ground swell, I saw this boat full of women and children at right angles—not parallel—to the ship's length and down near the ship's keel—or, to be exact, on the lower part of the ship's bilge.

At this point, Mr. Adams drew a picture of the situation he describes, right in the middle of the page. This picture has been carefully reconstructed and appears overleaf. Some minor details of the geometry of the ship in the Adams drawing have been corrected, but other than that it is a faithful reproduction. [See the original drawing in Appendix C.]

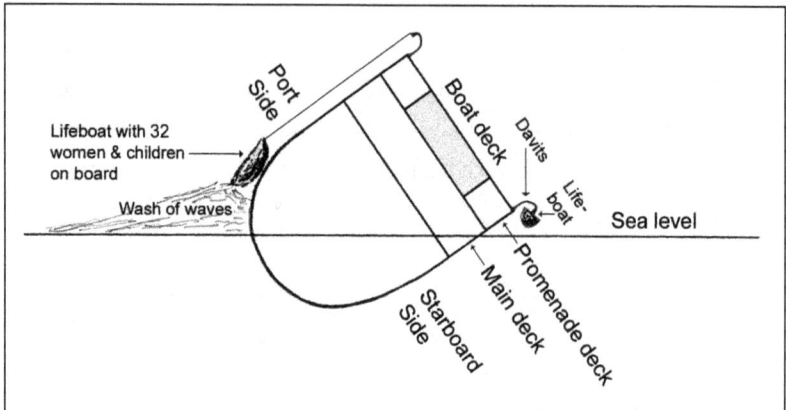

Original drawing by William P. Adams, courtesy of Ramon Jackson

The *Vestris* as she was about to sink

Adapted from a drawing done on November 20, 1928, by one of the *Vestris* survivors, William P. Adams. He drew this in a letter to Judson G. Jackson to illustrate the aspect of the *Vestris* as she was about to go under. Judson was the son of the Jacksons who perished in the disaster. The drawing shows one of the lifeboats with women and children on board, trapped against the hull of the ship. This was one of the lifeboats that was lost with all hands. The drawing shows the *Vestris* listing at about a fifty-five degree angle to starboard.

Adams continues his narrative:

> Several persons were standing on the side of the ship trying to push this boat into the water. Then our boat sank into the [trough] of a wave and the next time I got on top—and the ship also rose on a swell—that boat was gone and the ship had rolled onto her side and filled and sank in two or three minutes.
>
> Last summer I broke my arm and macerated my shoulder, and when I got on the *American Shipper* [after having been rescued], I saw a sailor with his arm in a sling and in fellow sympathy I got to talking with him and asked what had happened to his arm. He said he was in charge of the boat with the women

in it I have spoken of and as they were trying to work the boat down into the water a cargo boom or a davit—something very heavy—fell down on the boat and crushed it and in passing struck his shoulder (his forearm was later amputated). The occupants, more or less injured I assume, were thrown in the water alongside the hull of the ship and many must have drowned at that time and no doubt the rest were sucked under when the ship sank a few minutes after.

The boat next to ours was *very* leaky and filled and sank very soon after it was launched full of men.[12]

Dr. Groman gave this report:

§Apparently some of the officers of the crew had been trying to persuade the captain to let down the lifeboats. There was a bit of talk going on to this effect. Finally I heard the captain shout, "Go ahead, you might as well let them down."

There was a rush for the lifeboats. Everyone on board must have been aroused to the danger by this time. The crew started to loosen the first lifeboat. They could not unfasten it. Pulleys, ropes, and everything else became tangled and caught. Nothing seemed to work correctly. It was into this boat that they loaded the babies, accompanied by one man. It took them hours to get this boat halfway down the side of the ships. It would get caught on the port holes, doors for unloading freight, and whatever else was extending.

Suddenly a boom broke above the pulleys. It fell with a terrific thud into the center of the lifeboat, splitting it in half. All of the children fell into the sea and were soon swallowed up by the waves. The man in the boat had his shoulder crushed, and with a dead baby in his arms he dropped unconscious into the water.

Clearly, the man with the crushed shoulder and the one Mr. Adams met on the *American Shipper* are one and the same. Dr. Groman's story continues:

There was no panic on board ship. Everyone seemed quiet, though many especially the women, were weeping. Before their babies had been placed into the lifeboat they had given them crushed bananas, the only food available; it was almost poison, but the children had to have something to eat or they would have starved.

One by one the boats were lowered. It was less than 15 minutes before the *Vestris* sank that my friend and I entered one of the boats. It had a crew of colored men, with the exception of five whites. We dressed the boat perfect—that is, we had equal weight on each side—and then proceeded to row away.

The *Vestris* went down a few minutes later and we drifted into the night. We did not know which way to head and therefore drifted a great deal, also reserving our energy. We knew that the *Vestris* had drifted a long way from where she had sent her last SOS and so we tried to steer into her original position.[13]

Fireman Joshua Ford spoke eloquently of their predicament:

It is one thing to see us here now, but another when the sea opened its angry mouth and said, "I am ready for you." There we stood when the vessel listed way over. We looked to the east. We looked to the west. We looked all ways. There was no way to save us. There was the water waiting to take us, and we knew we were waiting for our deaths.[14]

Wallace M. Sinclair of Bound Brook, New Jersey, the South American representative of three American manufacturing concerns, gave this version of Monday morning's events:

I awoke Sunday morning to find a reasonably stiff breeze, but nothing untoward. By noon both sea and wind had picked up somewhat, and they continued steadily. But at no time would I class the weather as more than a reasonably stiff coastal storm. I heard the testimony of [Mr. Puppe] as to the lifeboats, and I should say that it was substantially correct.

A ship might list on account of the seas encountered, but this ship was slow to recover. She was not "alive." I did not, however, notice at any time anything that might have been interpreted as a shifting of the cargo.

On Monday morning there was a 30-degree list. I walked forward and saw members of the crew jettisoning the cargo in the forward hold. There were ten or twelve men lifting cases out of the hold by main force. They did not have any tackle. I watched them once for 30 minutes, and during that time they did not throw more than six cases overboard, for which I should say the total weight could not have been more than two tons.

I noticed that the forward well deck of the vessel was going farther under water. I thought then that the ship was doomed.

The thing that impressed me most was the lack of any intelligent plan of lowering the lifeboats or getting the passengers in them. I heard no order given to don life belts, and so far as I know the passengers did it on their own initiative. I was surprised as the list of the vessel increased—and consequently the difficulty of getting the lifeboats lowered over the side of the ship which was swinging farther and farther up—that the lowering of the boats was delayed so long.[15]

Paul Dana

Paul Dana was the South American representative of RCA (Radio Corporation of America) who was on a business trip to Buenos Aires. He gave this interview to Associated Press reporter Lorena Hickok, picking up where he left off on page 26:

My cabin was on the top deck, starboard side. When I woke up at eight o'clock Monday morning, there was the water—right on a level with my windows. There was water sloshing around on the floor of my cabin, too.

My steward came in, groaning, with a badly wrenched shoulder. There was apparently a leak, he said, the hold was filling with water, and he had been bailing—with a bucket—all night. The cargo had shifted too, and the ship had tipped clear over until the water was on a level with my windows on the top deck. I dressed as quickly as I could, ate the banana that was all my steward could get for my breakfast, and went up on the bow port deck.

The crew was throwing the cargo overboard—all the stuff they could handle. It looked like bales of cloth and such stuff. Their derricks weren't working, so they couldn't get rid of the heavier things, like automobiles. A little crowd of men stood up there, leaning over the railing and watching the crew throw the stuff overboard. Some of them had come out with their life preservers on. They looked badly scared. Although I had dressed myself that morning with the idea that I'd probably have to leave the ship before the day was over, I tried not to take such a dark view of the thing. Some of us did what we could to cheer the more timid ones—told them this was just a precaution and so on.[16]

The British Board of Trade inquiry into the disaster

Several members of the crew who survived the shipwreck testified before the Board of Trade inquiry into the loss of the *Vestris*, which was held in Liverpool, England, April–July 1929. This inquiry was presided over by Butler Aspinall, the Wreck Commissioner of the Board of Trade. Most of the questioning was done by an attorney for the Board of Trade named W. N. Raeburn. Among those testifying was Thomas Robinson, a first-class bedroom steward. He helped try to lower lifeboat Nos. 4 and 6.

I crawled on my hands and knees, and looking down I saw a steward named Evans going down with a child in his arms. There were several ladies in the boat. I did not see what happened to either boat afterwards. [Then] I was ordered to go to the starboard side, where a locker had fallen on one of the reels. We got it clear. Going again to the port side, I worked at number 2 boat. But we did not get it clear. I was almost immediately ordered back to the starboard side.

As I was going to the starboard side I heard the order to abandon ship, but I do not know who gave it.[17]

Thomas Connor, head waiter in the first-class dining room, stated that some of the crew had life belts on, and he got one for himself from one of the passengers' rooms. Later, he helped in the lowering of No. 8 boat, and then went to No. 10.[18]

§ One who had been going to sea for twenty-seven years was James McCulloch, a laundryman on the *Vestris*, who said he had made three previous trips in her. He said that on previous voyages he noticed water coming in through the port in the cabin and also through the half doors in the cross alleyway. He was in charge of a washing machine, and this also leaked. He had made reports about [as] something that was left to do, but it was done in the end.

Speaking of the last voyage, McCulloch said there was a steward in the ash house "up to his chin" in water early on Monday morning, trying to clear the scupper. He succeeded, but water kept coming in as the ship rolled.

Through my wife[†] being frightened—and, as a matter of fact, I had the wind up myself[‡]—I went up onto the boat deck. When I saw she was in such a position as she was, I said to myself, *This ship is finished.* I went to fetch my wife, but it took me all my time to go along the deck. I told my wife to come up and to bring what clothes she needed as she (the *Vestris*) was about to turn over. I helped

---

[†] McCulloch's wife was evidently Margaret McCulloch, listed as a laundress on the crew of the *Vestris*. She did not survive.
[‡] "Had the wind up": to become alarmed or nervous, agitated.

with the number 4 and 6 boats, and helped to put women and children into them. I put my wife and a stewardess into number 4. I got into number 4 myself, by walking down the ladder.[19]

The boatswain on the *Vestris*, Warwick Roberts, gave some rather revealing testimony at the British Board of Trade inquiry.

§ He said that he was having dinner about seven o'clock Sunday evening when the ship "appeared to strike something, and went over on her beam ends." She listed about thirty degrees, he estimated, and lifted "only slightly" but never got upright again. He said, "As soon as I could I extricated myself and went up on the well deck." He added that he met water on his way up, and on the well deck he found "a very heavy sea running." He finally went down to the crew's quarters to see how everything was down there and found a motor car (one of the three carried on the vessel) had gone right through.

The watch was changed at eight o'clock in the evening. Roberts said he had difficulty getting men for the watch. "They complained they had no dry clothes, and that they were injured, but I only saw one man injured." Eventually the watch was changed.

There was water in the starboard bunker at this time, most of which was coming through the scuppers in the cross alleyway. Only a little was squirting through the doors. There was also water in the alleyway leading to the starboard bunker.

During the whole of the night, Roberts continued, he was up and down the cross alleyway and found the water increasing. He could hear water washing about in the starboard bunker between three and four on Monday morning. He examined the [aft stowage locker] and found water there, but he had no idea where it was coming from. He reported this to the first officer.

At six o'clock in the morning, I thought the ship was doomed. I went down to the chain gang to cheer up the men who were bailing, and then helped move the car from the number 1 hatch. I went next to the boat deck, my boat being number 3.

Here, on the instructions of Captain Carey and the chief officer, I assisted with several boats and finally got number 3 into the water after much difficulty. I was thrown forward into it from a fish-plate. There were three stewards and a boy in the boat at this time. They had only one oar left, and the rudder was gone.[20]

The *Time* magazine article said, "It was 2:30 P.M. when the *Vestris* lurched on a billow to starboard and rolled under with a gulp, in froth and spume and reeling eddy. A few men, the last on board, sprinted across her horizontal side and dived. Captain Carey watched them, clung to his bridge."[21]

The fate of Captain Carey

There is some controversy concerning what happened to Captain Carey. Ernest Carpenter, second cook on the *Vestris*, said, "I was one of the last to leave the vessel. I saw Captain Carey standing near the stern, and then he jumped overboard. He had no life belt, and it is fairly well believed among the crew that he drowned."[22]

Second Officer Leslie Watson gave this account:

I jumped to the promenade deck rail and then to the next deck with Captain Carey, and we walked to [lifeboat] number 8, which was lying on her keel. I could not use the releasing gear because the boat was not sufficiently water-borne, it was not floating. I attempted to cut the falls. The ship sank then, and I was carried down by the suction, holding onto Captain Carey. . . . He never [surfaced]. When I came to the surface, I did not see number 8 boat, but there were several persons about, all apparently conscious and clinging to wreckage, waiting for a boat to pick them up.[23]

Passenger William W. Davies of New York told of the captain's last moments before the *Vestris* disappeared. He saw Captain Carey standing on the boat deck, worn and haggard from lack of sleep.

Many persons were still on board clinging to the rails on the deck below, and the captain called, "Good-bye, all." Then he went down with his ship.[24]

Lloyd R. Ricketts of Wirt, Oklahoma, said he was the last person before the captain and the wireless operator to leave the *Vestris*. He said that after he had left the boat he saw Captain Carey jump overboard. He said he could see him distinctly and asserted that the commander of the *Vestris* was not wearing a life belt.[25]

Harold Vowles, a steward, was asked if he had seen Captain Carey when the *Vestris* went down. He replied that the captain was standing on the bridge with a bugle when the vessel disappeared beneath the water.[26]

Passenger Edward Miles Walcott, of Georgetown, British Guiana, said, "Although I had a life belt on, the suction pulled me seven to eight feet under water. When I bobbed up, Captain Carey was alongside me, only a few feet away. I saw him distinctly and recognized his features. He did not have a life belt and sank again almost immediately."[27]

Second Steward Alfred Duncan said that after all the *Vestris*'s passengers were assembled on deck to take to the boats, he went to Captain Carey, who was on the bridge, and asked if there were any further orders.

> He was wearing a greatcoat, but no life belt, and seemed quite calm when I spoke to him. "I think that all the passengers will be saved," Captain Carey said. "Get your life belt on. Pay no attention to me."
>
> Those were the last words I heard him say. I put my life belt on, and by that time the water was one-third over the deck on the starboard side and the ship began turning over slowly on her starboard side. I jumped into the water, but Captain didn't follow me. I don't know what happened to him after that, as the *Vestris* went over on her beam ends, and after a minute or so she wallowed twice and plunged down to Davy Jones' Locker.[28]

Carlos Quiros, previously quoted, said, "Captain Carey remained standing at his post until the end, giving last-minute orders. As Davies and I jumped last off the ship, Captain Carey walked calmly to the *Vestris*'s railside. That was the last I saw of our captain."[29]

These accounts justify the current view that eyewitness testimony is unreliable. All they agree on is that Carey had no life belt when the ship sank and did not survive. They have him on the stern, by lifeboat 8, on the boat deck, and on the bridge. Some say he jumped off, others that he went down standing there. One said he never surfaced after the ship went down, others say that he did.

The *Vestris* lost beneath the waves, the survivors were left to deal with their lot, alone now in the vast expanse of the Atlantic Ocean.

≈ ≈

Disposition of the lifeboats	
Number 1	Launched successfully; rescued by the *American Shipper*.
Number 2	Lost upon launching; broken by falling davit.
Number 3	Launched successfully; rescued by the *American Shipper*.
Number 4	Went down with the *Vestris*, still on the davits. One end fell and spilled all occupants into the sea prior to the *Vestris* sinking.
Number 5	Launched successfully; rescued by the *American Shipper*.
Number 6	Went down with the *Vestris*, still on the davits. All aboard were lost (reportedly held thirty-two women and children).
Number 7	Launched successfully; rescued by the *Myriam*.
Number 8	Launched but was punctured and subsequently capsized. Four of its occupants were rescued by the USS *Wyoming* clinging to the capsized lifeboat; they swam to the *Wyoming*.
Number 9	Never launched.
Number 10	Launched successfully; rescued by the *American Shipper*.
Number 11	Launched successfully; rescued by the *Myriam*.
Number 12	Sank soon after launching.
Number 13	Launched successfully; rescued by the *Berlin*.
Number 14	Launched successfully; rescued by the *American Shipper*.

Radio messages from the stricken *Vestris*

On November 19, 1928, Attorney Tuttle convened the hearing to get testimony about radio messages exchanged during the foundering of the *Vestris*. The first witness was Arthur J. Costigan, who produced the log books of all the Radio Corporation's stations on the Atlantic Coast—at Chatham, Massachusetts; East Moriches, Long Island, New York; New London, Connecticut; Bush Terminal, Brooklyn, New York; and Tuckerton, New Jersey—that would be likely to receive such messages.

§ "Will you tell me," Attorney Tuttle asked, "whether there is any message in any of these logs purporting to have been sent on Sunday, November 11, the day after the *Vestris* sailed from New York and the day before she foundered, and dealing in any way with the steamship *Vestris*?"

"There is not," Costigan retorted. So saying, he put to rest speculations by some of the ship's passengers that Captain Carey had sent out an SOS on Sunday and then cancelled it.

Costigan then reported the entries recorded in the radio log of the Tuckerton station, which was nearest the scene of the sinking and therefore handled more messages dealing with the *Vestris* than the other stations. He testified:

> At ten minutes past nine on the morning of Nov. 12 the *Vestris* called Tuckerton and said, "We are heeling over, and may need assistance. Please broadcast the following to all ships and stations: Keep close watch for distress calls from SS *Vestris*, who may need immediate help." That broadcast was completed by Tuckerton at 9:18 A.M. [This was the CQ message.]

> At 9:58 A.M. Nov. 12, *Vestris* sends SOS followed by, "We are heeling over 32 degrees on our starboard side and need immediate assistance." Two minutes later it was broadcast to all ships by Tuckerton on a more powerful transmitter.

Additional radio messages testified to by Mr. Costigan follow:

10:02 A.M.: Tuckerton called the *Vestris* and [requested], "Give me your approximate position, please."

10:08 A.M.: The *Vestris* called Tuckerton in reply, "We are in approximately latitude 37.15 north, longitude 71.08 west."

10:15 A.M.: That information was broadcast by Tuckerton to all ships, with the request that any ship in the vicinity of latitude 37.15 north, longitude 71.08 west, please advise Tuckerton immediately. [Ed. note: This position was corrected later from radio bearings.]

10:20 A.M.: The steamer *Santa Barbara* advised its position and says that it can get to the *Vestris* by 7:30 P.M.

10:30 A.M.: The *Vestris* again sends out SOS, followed by "Urgent help needed. We are already 32 degrees list to starboard and want help immediately."

10:40 A.M.: The SS *Ohio Maru* advised the *Vestris* she is about 135 miles from it and can render help if wanted. The *Vestris* replies and says: "We are getting worse. Decks all under water. Ship lying on beam end. Impossible to proceed."

11:00 A.M.: The *Vestris* asks, "Oh, please come to our assistance."

11:10 A.M.: Tuckerton is busy collecting position reports from all ships in the vicinity in an effort to ascertain which is the closest.

11:20 A.M.: The *Ohio Maru* tells the *Vestris* unable to go to him as they have full cargo load. [Costigan commented:] That, by the way, was disproved some time later.

11:40 A.M.: The *Vestris* is getting radio bearings from Cape May and NJW—I believe that is Amagansett [Long Island, New York]; I am not certain of it.

12:00 noon: Tuckerton broadcasts to all ships "Stop sending SOS."

12:15 P.M.: The *Vestris* called Tuckerton and said, "It is bad enough in this situation, old man—let alone position ship is in. Can hardly stay and receive." Tuckerton replies "O. K. Have notified naval station in the Brooklyn Navy Yard, and they have dispatched a cutter to your help."

12:30 P.M.: No change in situation.

12:45 P.M.: The *Santa Barbara* says [it] will go to the assistance of the *Vestris*, and that [it] has several good radio bearings on the [*Vestris*] and is due to get there about 7:30 P.M.

1:03 P.M.: The *Vestris* is busy getting radio bearings from the naval compass stations ashore.

1:10 P.M.: The *Vestris* called Tuckerton and says, "Our steam is gone. Power getting low. Will use coil. Please listen for same." Tuckerton replies "O. K."

1:15 P.M.: The *Vestris* tests out its coil, that is, its emergency transmitter, and says that [the] main [*Vestris*] transmitter is dead now. No power from engine room.

1:20 P.M.: Tuckerton knows that the coil signals, or emergency set signals, from the *Vestris* are O.K., and that the *Vestris* is then making signals for the naval compass station, Cape May, to secure bearings.

1:27 P.M.: Tuckerton broadcasts to all ships radio bearing of the *Vestris* as given by the naval compass station at Cape May for the benefit of all ships in the vicinity.

1:29 P.M.: Notation to the effect that the coil signals of the *Vestris* fade out and Tuckerton is unable to hear them.
Again the signal comes in faintly, saying, "Going to abandon ship in few minutes. Getting lifeboats ready now."

1:30 P.M.: *Vestris* announces, "Abandoning ship now." A few minutes later. "So long WSC." [Ed. note: WSC were the call letters of the Tuckerton, New Jersey, radio station.]

1:35 P.M.: Tuckerton informs the naval stations that the *Vestris* has been abandoned, and suggests the resumption of commercial [radio] traffic, which has been stopped during the distress period.[30]

All of these messages from the *Vestris* were presumably sent by the chief radio operator, Michael Joseph O'Loughlin. Despite reports that O'Loughlin went down with the ship still at the wireless set, he apparently was lost trying to get on board a lifeboat according to testimony from survivors Second Operator James MacDonald and Third Operator Charles Verchere.

Further testimony by Costigan reveals the crucial radio bearings obtained while the *Vestris* was sinking. These records are from the Chatham, Massachusetts, radio log.

> 11:15 A.M.: The *Vestris* again obtained radio bearings from ashore.
> 11:30 A.M.: The *Giorgio Ohlsen* relays to the *Vestris* the radio bearings as ascertained by NJW: "Fire Island [Long Island] 146 degrees, Sandy Hook [New Jersey] 141 degrees, Mantoloking [New Jersey] 128 degrees, and Amagansett [Long Island, New York] 165 degrees."
> 11:35 A.M.: The *Vestris* gives to the *Cedric* his latest radio bearings: Cape May [New Jersey] 115.5 degrees, Cape Henlopen [Delaware] 110 degrees, Bethany Beach [Delaware] 104 degrees, [as of] 11:13 A.M.[31]

These radio bearings are important since they contradict the theory that Captain Carey was depending only on dead reckoning for the positions he gave in his SOS messages. The position given by these radio bearings, 37 degrees 35 minutes north, 71 degrees 8 minutes west, was evidently as accurate as possible using the technology available in 1928.

The rescue ships proceeded to that location, oblivious to the drift that would have occurred since the *Vestris* sank there—in the middle of the Gulf Stream.

Fred Hanson / New York *Daily News*

The *Vestris* listing badly to starboard

This photograph, taken on November 12, 1928, the day the *Vestris* sank with the loss of 111 lives, shows the ship listing about 29 degrees. The snapshot was taken by Fred Hanson, an assistant pastry chef on the ship's crew, using a folding Kodak camera he bought the day before the *Vestris* set sail. He took this picture at 12:45 P.M. from the port side, looking aft; hence the starboard side is on the left. One of the lifeboats can be seen hanging in its davits near the top of the picture.

Note: Unless otherwise identified, all photos of the sinking *Vestris* were taken on Monday, November 12, 1928.

Abandoning ship

One of the lifeboats is shown leaving the starboard side of the *Vestris*. This photo has been straightened to show a level horizon, as it should. The list of the ship is close to 35 degrees in this picture. Photo taken by a passenger.

Fred Hanson / New York *Daily News*

Launching a lifeboat from the *Vestris* at 12:30 P.M.

Lifeboat 1, which was rescued by the *American Shipper*. The man is looking down from the Promenade Deck; Fred Hanson took the picture from the Boat Deck, November 12, 1928.

Fred Hanson / Library of Congress

Launching lifeboats

This picture—probably taken within minutes of the one in the frontispiece—was taken at a true vertical, as can be seen by the orientation of the people on the deck. The list of the *Vestris* here is twenty-eight degrees. Small differences in list may be attributed to the ship rolling as it passed over the swells, which were said to have been waves twelve to eighteen feet high from crest to trough. Some port-side lifeboats are still hanging from their davits. The unfortunate steward George Hogg with his broken arms is leaning against the wall at the far left, just as he is in the frontispiece.

The badly listing *Vestris*

This grainy photograph, widely disseminated on the Internet, is likely number five of Fred Hanson's photos (per the British Board of Trade Inquiry into the *Vestris* Disaster), which he took at 9:15 A.M. Monday morning. He described it as "taken from the half-deck, near boats number 15 and number 14. We see number 14 inboard on the port side."[32] The list of the *Vestris* in this picture—as defined by the angle between the mast and the horizon—is 29 degrees, in close agreement with the list angle in the photos taken by Hanson about nine o'clock that morning. The picture was first published on November 15, 1928, in the *Baltimore News*. *Wikimedia* asserts that this photo is now in the public domain; its copyright expired and was not renewed.

Some minor damage is evident on the lower portions of this lifeboat, and the overlapping clapboard "clinker-built" construction is obvious. This is what made it so hard to lower the port-side lifeboats, as the clapboards kept getting caught on the hull plates of the *Vestris*.

Chapter Six

At the Mercy of the Sea

IF PANDEMONIUM characterized the launching of the lifeboats as the *Vestris* was sinking, desperation was what soon followed on its heels. Nine lifeboats drifted in the still-heaving seas, although one of them, No. 8, would soon sink. Most of the lifeboats were without any or enough supplies. Some had workable flares. Few apparently had any potable water or edible food on board. The boats were made up of two from the port side—Nos. 10 and 14—and six from the starboard side—Nos. 1, 3, 5, 7, 11, and 13 (this was one time when 13 was a lucky number).

No. 1 was a motorboat that nobody on board knew how to operate. Each lifeboat carried from five to forty-seven survivors. There were also survivors who hung on to bits of wreckage and one man who survived with only his life belt for support.

They were more than 200 miles from the nearest land and had no means of communicating with the outside world. The weather was not favorable—starry skies one moment and showers or hail the next. Waves continually threatened to upset or breach the boats.

As mentioned previously, the *Vestris* carried fourteen lifeboats. Four were carried on the poop deck (a deck that forms the roof of a cabin built in the rear, or "aft," part of a ship's superstructure). Ten were carried under davits on the top deck. The five on the port side had even numbers, those on the starboard side, odd numbers.

This chapter is divided into four major sections: one for each of the four ships that rescued the survivors. These four sections are further subdivided to show which lifeboats were picked up by those ships. For each of these lifeboats, we relate first-person accounts from its survivors, as they were available to us.

The *American Shipper*: First of the rescue ships

The *American Shipper* was first on the rescue scene thanks to the navigating skills of Captain Cumings, the ship's master. This picture was distributed by Pacific & Atlantic Photos and published in *The Rockford* (IL) *Daily Register -Gazette* issue for November 15, 1928, on page 6.

Rescues by the *American Shipper*

The *American Shipper* was a cargo vessel operated by the American Merchant Line. She was on her way to New York City when she was called upon to play a most important role in the rescue operation. Her navigating officer, E. A. Ohman, made this statement in a radio address on November 14, 1928:

> On Monday [November 12] our steamer, the *American Shipper*, was well on her way to her home port, and we expected to be anchored in New York Harbor by midnight Monday. But at six minutes past eleven in the forenoon on Monday our wireless man reported the receipt of an urgent distress message from the *Vestris*. This message gave the latitude and longitude of the *Vestris*, indicating that she was approximately 145 miles south of us.
>
> Immediately on receipt of the message Captain Cumings turned the ship around and headed directly for the position given, at the same time sending word to the chief engineer to give the ship full power under forced draft. At about 7:30 P.M. we arrived at one of the positions given to us by shore radio stations, but there was no sign of a disaster around there, and as the position given by the *Vestris* was approximately thirty miles south, we proceeded there, but again we were disappointed, as the only thing we encountered was another steamer search[ing].
>
> We kept going farther to the south from 11:08 P.M. until 2 A.M. with everyone on the lookout for the boats or wreckage, but at 2 A.M. Captain Cumings, after carefully checking on the various positions given and making allowance for the drift of the Gulf Stream and the wind, figured that the lifeboats would be farther to the east. This conclusion proved to be correct, for at 3:40 A.M. I, who was in the crow's nest, sighted a red flare almost dead ahead, and at 4:05 A.M. we arrived alongside the first lifeboat.[1]

Apparently Captain Cumings was the only one out there who realized that the debris and lifeboats from the *Vestris* would not stay at the location given in the SOS but would drift with the Gulf Stream to the east. The lifeboats were found about fifty nautical miles east of the SOS position.

Lifeboat No. 5

No. 5 was the first to be picked up by the *American Shipper* at just after four o'clock Tuesday morning. One of the passengers in lifeboat 5 was John Santana, who testified at the Tuttle hearings.

When the ship sank, he was thrown into the water and found himself floating, supported only by his life belt.[2]

> [Just prior to that] I found myself on the topside, alone, save for Captain Carey. He was wearing a heavy top coat, but no life belt. The ship began sinking fast. You could feel her going down beneath you.
>
> When we went down it seemed a huge wave was pouring over us, and the captain and I were in the water. We were drawn down by the suction. I came up and reached for a floating box. I didn't see the captain again. I don't know how long I floated, but I finally [found] lifeboat 5.
>
> I had been swimming after the boat for some time, but the Negro members of the crew rowed away.
>
> Other people were begging to get aboard, too, but the Negroes didn't seem to want any more aboard. [Then] Chief Engineer Adams made them stop and we swam to the boat and were taken in.[3]

Santana said the persons in lifeboat 5 included only five passengers. The boat was leaking badly, and they bailed all night until they were picked up by the *American Shipper* Tuesday morning.[4]

"The water was all right," said Santana. "It was warm enough and afterward the people on the *American Shipper* told me that the temperature was about 78. But the air was very cold. I was cold all night, and when it rained it was even worse.

"When we got aboard the *American Shipper*, they treated us great. They took us below and dried out our clothes and gave us something to eat. Nothing could have been better than the way they treated us."[5]

Steward George Hogg, forty-one, of London, England, had both his arms broken when a davit broke and fell on him as he was trying to help launch a lifeboat. He can be seen in the frontispiece on the deck of the badly listing *Vestris*, leaning against the ship's super-structure and holding his crushed arms in front of him. He was picked up by lifeboat 5 and rescued by the *American Shipper*.

§ Two others known to have been on lifeboat 5 were Thomas Connor, head waiter in the first-class cabin, and Third Officer George Herbert Welland, who was in charge of the lifeboat. At the Tuttle hearings, Connor said he first had gotten into lifeboat 8. The men lowering the boat let go of the ropes holding one end and that end fell, tipping the boat endwise. He was thrown into the sea.

Connor was picked up by lifeboat 5, of which a crewman, a lamp trimmer, was in charge. He said he "was out, was exhausted," but recovered an hour later. Subsequently they picked up the third mate [Third Officer Welland] floating by on wreckage from the deck-house, "more dead than alive"; a man [steward George Hogg] with both arms broken; a woman passenger, also "more dead than alive from the look of her"; and two male passengers (the three passengers were not identified).

The lamp trimmer was probably Charles Harris, who was listed as "in charge of lamps." The term "third mate" is synonymous with "third officer," so this was most likely Third Officer Welland. How he rose from having been "more dead than alive" to being "in charge" of the lifeboat is a mystery.

The boat was fully equipped according to Connor, and after rowing a while the sail was hoisted. It was not necessary to bail more than about half the time and it did not leak more than "any lifeboat until she swelled."[6]

Lifeboat No. 1

The second lifeboat to be liberated by the *American Shipper* was No. 1, probably around 4:30 in the morning. Nobody was "in charge," but occupants included Anne DeVore of Dayton, Ohio, and her wire-haired terrier, Speedway Lady; William Davies; Fred Hanson; ship's carpenter Gustav Wohld; and fifteen others.

Anne DeVore's story is among the more fascinating of this sad saga. Anne and her husband Earl, an Indy car driver, were in lifeboat No. 8, which had been holed in launching and the hole repaired with a piece of tin. With them in the boat were Norman K. Batten—another race car driver—and his wife, Marion, of Altoona, Pennsylvania. They would not be in that boat for long.

§ Despairing of keeping the boat afloat, although the women joined the men in desperate bailing, Anne DeVore said she had hailed another lifeboat carrying only four members of the crew. They came near and she leapt with her dog to that boat. She said the men refused to wait for her husband and the others to follow.

> We had nothing else to do but to take [number 8]. It was overcrowded, and as soon as it got into the water it began leaking. It also shipped a lot of water and was filling fast. Number 1 lifeboat with four Negro seamen in it came by. We called to them to come near us and take some of us aboard. They drew near and my husband urged me to jump. I did so holding Speedway Lady in my arms, and had no sooner gotten aboard than they pulled away. A sea rose between us and when I could again see number 8, it was capsized and everybody was in the water. I begged the men to go back and save my husband and friends, but they absolutely refused to do it.
>
> As the men rowed away, I heard my husband shout from the other boat, "Help! We are sinking!"

They said they must save those swimming around in the sea, and they did not seem to understand English. They responded every time a West Indian member of the crew called to them in Portuguese or in Spanish, but apparently did not understand the cries for help in English that were issuing from all around.[7]

Anne DeVore, who saved her wire-haired terrier by holding her through the long night as she was tossed about in the boat, paused and her eyes filled with tears.

"Those men were no better than murderers," she said.[8]

CAPTAIN FREDERIK SORENSEN, a second-class passenger, gave this slightly different account of the same episode, in which he said that he left the ship about one o'clock Monday afternoon in lifeboat No. 8 with about forty or fifty other persons, mostly women and children.

The boat sank after we had been afloat hardly more than half an hour and only four of us were saved. There was a hole as big as that [he indicated about six inches in diameter] in the boat when it was put over the side and some of the crew had nailed a piece of tin over it with four nails. That boat ought never to have been lowered [into the water].

There wasn't an officer in the boat either, only some of the crew and passengers. We were about 15 percent men, the rest women and children. We all took turns rowing and most of us had to bail furiously with our hats and everything else we had to keep afloat as long as we did.

The boat soon filled with water up to the gunwales and was swamped. Then lifeboat 1 passed us about a hundred feet away, with only about eight persons in it: one woman, Anne DeVore, and the rest colored waiters and members of the crew. Suddenly I heard Mrs. DeVore scream that she heard her husband's voice among those of us in the capsized boat who were still clinging to its sides.

"For God's sake stop and get them!" Mrs. DeVore cried, but the men kept on rowing and wouldn't pay any attention, as they were afraid there were so many of us that we might sink them. They were murderers—nothing else. "We can't take 'em—they're too many!" they told her.

Just about that time I was washed out by the sea. Women all about me in the water were wailing, and one whom I knew as a Mrs. Raphael, cried "Oh, my God! The lifeboat's leaving us!" and went down. That was the last I saw of her.[9]

Captain Sorensen retold his account at a later date, adding some more information:

Number 8 boat began to ship more water. There began to be some excitement on board. The men shouted "Bail" and began to bail with everything they had. I asked them to keep calm. At that time I hoped that number 8 boat would stay afloat on her air tanks. But the water gained on us so rapidly that I soon knew she was going to be swamped, but I didn't want to frighten those in the craft by telling them so.

When we saw the water gaining like that we started to make for some of the other boats. But as soon as we reached any other boat it would pull away from us and we had a hard time to get down to them. Finally we got to number 1 boat. When we reached number 1 boat our own boat was nearly filled.

There were only eight men in number 1 boat. The women in our boat were standing up in the water. One of them I think was holding up her child. They called out "Please come and save us!" Number 1 boat just stayed there and the people in it just watched us, all except the woman, Mrs. DeVore. I saw her moving her arms and I guess she was talking to the men in that boat.

Just then number 8 boat became filled with water and swamped. Just as the sides went beneath the water and the water came up on us I heard Mrs. Raphael scream: "My God the boat is leaving us."

When number 1 boat saw that we were being swamped it pulled away. This is something I can never forget and it makes me feel pretty bitter against that crew. There were only eight men in that number 1 boat and Mrs. DeVore. They had a capacity for sixty people and they could at least have taken of all the women and the children from number 8 boat and have let the rest of us hang on to their lifelines.

Earl DeVore was on our boat. His wife heard his voice calling across the water, and as I learned afterwards she begged and pleaded with the cowards on number 1 boat to come to our rescue. Instead of doing so they pulled away and left the people on number 8 boat to drown. I will never forget my feelings as I saw number 1 boat pull away. I was wishing for a gun and gosh if I had had one I would have used it. If one of those cowards in number 1 boat had been shot the others would have come along.[10]

We continue with an account from William Wills Davies, New York representative for the Buenos Aires newspaper *La Nación*:

§ "I saw Captain Carey standing on the boat deck, looking tired and worn out. Many of us were still on board clinging to the rails on the deck below and [we heard] the captain call 'Good-bye, all.' "

The *Vestris* suddenly rolled over on her side, and Davies found himself clinging to the rail post, which was now horizontal instead of vertical. His feet dangled as he hung on for dear life. Looking down, he saw the sea rushing toward him. He dropped with perhaps a dozen other passengers and was drawn far down. As he started to rise he tried to ward off debris in the water.

For a while he was afraid he would not make it to the surface and recalled hearing that drowning was supposed to be an easy way to die. But suddenly he burst through to the surface of the sea, and the fresh air he breathed in banished those thoughts. Floating debris of all sorts was everywhere. There were dozens of struggling persons all around him. And the *Vestris* had simply disappeared.

Davies at first clung to a chair, but gave that up quickly when he came upon what looked to be the top of a hatch. In the roiling sea he saw his friend, Mr. Koppe, who was also clinging to wreckage. People were swimming everywhere, but the boats were staying away, afraid of taking more people aboard than they could safely hold.

Davies saw a crewman lying on a piece of wood and holding a knife in one hand as though he meant to keep anyone from trying to share his precious log. Davies passed a capsized boat and yelled to another but got no reply. A third boat came near but the people on it told him to swim for it as they would not come near his hatch. They lifted him aboard with a boathook. There was one woman on this lifeboat, Mrs. DeVore. Davies praised two men in charge, George Hansel and T. Griffin, and told how they kept three or four crewmen at work rowing, though the crewmen were tired from lack of sleep and complained bitterly.

Davies said it was a terrible night. The wind would dry his clothes, and then a shower would soak them again. At times he was afraid that he might lose his mind.

About eight o'clock they saw a searchlight beam play across the sky, but it was a tantalizing apparition as it seemed to come no nearer. At nine o'clock they saw a faint light but it faded away. They had just used their last red flare when they saw the lights of the approaching *American Shipper* and were rescued.[11]

There was nobody named George Hansel on the *Vestris*; he most likely meant storekeeper George Amsdell. T. Griffin was Thomas Griffin, a waiter on the *Vestris*. Both survived the sinking.

Another person on board lifeboat 1 was Gustav Wohld, the ship's carpenter. He testified at the British Board of Trade Inquiry into the *Vestris* disaster:

Describing the launching of the lifeboats, Wohld said he heard Captain Carey give orders for the boats to be lowered, and he thereupon went up to No. 4 boat.

When asked what happened to that boat, he said it stuck between the boat and promenade decks, and the planks were damaged.

He then went with some of the crew, to help with the starboard boats. He helped to lower No. 3 and No. 5 boats on the starboard side, and then went forward to No. 1, of which he was in charge.

He got the boat into the water and away from the ship with two stewards, a waiter, and himself. He couldn't get back to the starboard side of the ship as it would have damaged the boat. They took up some people who were swimming in the water; finally there were about nineteen people in No. 1, including two passengers and a dog.

When asked if the motor worked, he said the motor was in good condition, but "we had no engineer in the boat, and I did not know how to work it myself."

Asked if the boat had all the equipment in it, he said [it] was in good condition and did not leak. In the end, added Wohld, they were picked up by the *American Shipper*.[12]

Lifeboat No. 3

Lifeboat 3 was the third to be picked up by the *American Shipper*, with Chief Officer Frank Johnson in charge. It was picked up about 5:15 A.M. Tuesday, close to fourteen hours after the *Vestris* sank.

Published reports state that when lifeboat 3 was launched it carried only eleven crew members, no passengers. The lifeboat did, however, pick up twenty-seven persons from the sea. When the *American Shipper* rescued her, there were forty-five persons on board the boat, making it one of the most heavily loaded of all.

One of those picked up by lifeboat 3 was Herbert C. W. Johnston, manager of an oil field in Trinidad, who gave this account:

The first two boats to be filled contained mostly women and children. One of them swung against the side of the ship and was holed as it was being lowered. The other capsized a minute after it struck the water. The passengers from both were spilled into the water, and as there were no other boats launched there was not one to pick them up.

It was obvious that no boats could be successfully launched from that side, so I went to the starboard side. I walked down the sloping side and swam to No. 3, which contained about a dozen Negroes. I tried to get them to row around to the port side of the *Vestris* but the sea was so rough we couldn't make much progress.

There was nobody in charge in the boat and all the Negroes were shouting commands, so when we met number 3 boat, which had about twenty people in it, I transferred to it.

As we rose on the crest of a wave I saw the *Vestris* heel over on her side. Then there was what seemed to be a puff of steam and she was gone. The boat I was in and two others rowed back to the scene, and we picked up about twenty-five people who were floating in life belts.

I saw one lifeboat go down after the ship, but those in it had life belts on, which floated them. I think that most of them were picked up. Our boat cruised around picking up people until we were loaded to the limit. We must have had forty-five on board by that time. There were still fifteen or twenty people clinging to wreckage in our neighborhood whom we couldn't save.

I saw eight boats get away safely, which should have accommodated 300 people easily, but one of the boats I saw had only eight people in it. There wasn't the slightest sign of panic. The passengers and crew behaved wonderfully.

We cruised around, calling to the other boats in an effort to keep together as darkness fell. From a swamped boat we picked up Chief Officer Johnson and he took charge. The seamanship of the Barbados crew members in our boat was splendid.

About eight in the evening we saw the searchlights of a rescue ship in the distance. A thunderstorm was blowing up and we got fearfully excited at the prospect of being picked up before it broke. We lit flares and hoped for the best, but about half an hour after we had first sighted them the beams of the searchlights passed out of sight.

It had started to rain by that time and we were all wet through and fearfully downcast. The sea was getting worse all the time. The outlook seemed to be pretty hopeless, but we opened a tin of biscuits and ate them, which helped some.

Every now and then through the night we thought we saw searchlights and lit flares to try and attract attention. Several times we mistook low stars for the lights of a ship. About three o'clock in the morning we distinctly saw the searchlights of a ship, but our flares went unanswered.

Not long after that the searchlights of the *American Shipper* came into view and came closer and closer. We only had two flares left, both of which had been damaged. Their caps were gone, but my experience in mining had taught me how to set them off in spite of that.

I took one of the flares in the palm of my hand and our last box of matches between my fingers, then struck a match right down from the side of the box into the powder of the flare. The *American Shipper* seemed to see it. At any rate, it kept coming nearer.

By this time the wind was blowing a gale and the sea was getting constantly rougher. We deliberated whether to risk capsizing by trying to row to the ship or drift until daylight in the hope of being picked up then.

We drifted for about an hour and a half more and then lit our last flare. The *American Shipper*'s searchlights picked us up again and held us from then on. As the ship approached, the sea kept getting worse, so we decided to risk rowing and turned toward it.

At last we came alongside. We were all fearfully cold and some of the passengers were so weak that the crew of the *American Shipper* had to hoist them up the ladder.

I want to give the warmest praise and thanks to the officers and crew of the *American Shipper* for the wonderful treatment they gave us. Captain Cumings deserves tremendous credit for his splendid seamanship, particularly the way he allowed for the current and drift. When we got aboard they had hot coffee and soup and hot blankets waiting for us, and after we had been warmed up and fed we turned in for a long sleep.[13]

Chief Officer Frank William Johnson, who was picked up by boat No. 3 and assumed charge of it, gave quite a bit of testimony before the British Board of Trade in England in June 1929, including this about the lifeboats:

When asked if passengers who were to be put into the starboard boats would have had to jump over and swim or be dragged to them, Johnson replied that they would have had to climb over the rails and jump into the water. Questioned if he knew if any attempt was made to get the passengers over to the starboard side, he said he didn't remember.

When asked if the Board of Trade requirement that the hooks of the lifeboats' releasing gear be lengthened were complied with, he said they were not, "but the American people were satisfied."

Johnson said he was sent to lower the starboard boats, and the last boat he had worked with was No. 1. Asked if having got his boats away he made an attempt to get up to the port side to see if everything was all right, he said, "I tried, but could not manage it."

When asked where No. 9 was when he left, Johnson replied, "Lying beside the ship, almost waterlogged by the falls not having been cut and the boat being upended."

He stated that it was not possible [toward the end] for people to get from one side of the ship to the other.[14]

Johnson was recalled for further testimony the next day. Attorney Butler Aspinall asked Johnson precisely what he had done with the lifeboat he was dealing with first. Johnson replied, "We got the falls cleared and let go the grips, and the boat swung out. We only had to lift one boat, No. 9."

He stated that some people got into No. 9 before it reached the water, but that he did not. He added that a number of the crew got into lifeboat 7 before it reached the water by climbing over the davit. He said he jumped into the water after he had lowered boat No. 1.

"Was there any reason for your jumping and not getting over the davit as the others had done?" asked Aspinall.

"Well, I lowered the No. 1 boat last. I looked around but could not see there was anything else to do, so I saw No. 9 boat lying about, and I jumped into the water and swam to that boat.

He said he was in lifeboat No. 9 when the *Vestris* sank.[15]

Another one who was on lifeboat 3 was Edward Miles Walcott, a shipping agent from Georgetown, British Guiana. This is his story as pulled from several sources:

> The lifeboats should have been lowered and the women and children put aboard them much sooner. The crew seemed to be green. They behaved coolly, but inefficiently. They didn't know how to launch the boats or how to handle them in the water.
>
> Another man and I were making our way along the sloping deck toward a lifeboat which was almost full when we saw two women, so we shoved them up into the boat. There wasn't room enough for us, and we didn't have a chance to get to another boat before the ship went down.
>
> Although I had a life belt on, the suction pulled me seven to eight feet under water. When I bobbed up, . . . there was a woman on the other side of me, supported by a life belt. Two oars came floating past, and I grabbed them and passed one of them over. After a while a plank came along, and we wedged it under the oars and made a sort of raft.

We were in the water about three-quarters of an hour when a boat [No. 3] came by with about twenty members of the crew, but no passengers in it. They stopped and picked me up. At first they were not going to take the woman, but I insisted.

They claimed that the boat was leaking and couldn't take any more. It was leaking, all right, but it could have held ten or twelve more. I told them that I had seen a woman floating not far away, but they wouldn't go back. Instead they rowed away.

There were four colored men aboard who behaved better than most of the white men. When it got cold at nightfall most of those in the boat stopped bailing and huddled under a tarpaulin. Three of the Negroes, a Frenchman, another white member of the crew, and I were the only ones who kept bailing or trying to handle the lifeboat. We could have lasted a number of hours longer I believe, although I was nearly exhausted when the *American Shipper* picked us up.[16]

In the Hoover hearings at the Custom House, Walcott told a slightly different version of the events when the *Vestris* went under.

Under questioning, he said he had waited until women and children were loaded before looking for a lifeboat for himself. At that time two women appeared "from nowhere" he said, and a third woman came in view trying to crawl to the upper, port side of the vessel. Walcott slid down the deck to help her, and as he was doing this the *Vestris* sank.

"I went down some distance," Walcott related, "and when I came up again I had the good fortune to find an oar floating nearby. Presently I saw the third woman near me and we shared the oar. We found a plank or two a little later that made a comfortable raft to float with."[17]

The woman he hooked up with was Mrs. Laurinda Moore of 21 Clinton Avenue, Jersey City. Still shaking from her experiences, the elderly stewardess from the *Vestris* sat on a sofa in the *American Shipper*, white-haired and bent at the waist, disinclined to talk.

"Oh, please don't ask me!" she said in reply to reporters pressing her for information. "The horrors are in my head."

She had attempted to comfort a woman on the *Vestris* who said she had just seen her husband hurled from a lifeboat into the raging waves of the roiling sea.[18]

"He will be picked up," Mrs. Moore assured, "keep a good heart."

She then saw a skinny woman holding a baby in her arms trying to climb into a lifeboat. Mrs. Moore helped the mother into a lifeboat and took the infant in her own arms.

As she was getting into another lifeboat with the baby, a rope holding the lifeboat in position broke or slipped out of the hands of the man who was holding it. Mrs. Moore and the baby were thrown into the water, and the baby could not be found.

Mrs. Moore floated with the help of her life belt for an hour and a half before she was picked up by No. 3 in which were five persons. What happened after that she couldn't recall, for she lay comatose in the bottom of the boat. They were in it all night with only some hardtack to eat before the *American Shipper* rescued them.[19]

This account omits the details of her watery alliance with Edward Walcott and the makeshift raft, and there were more than five persons in the boat she got into. It began with eleven crew members, and picked up a number more. In all likelihood these are just typical of the discrepancies found in eyewitness accounts.

Another who was rescued from lifeboat 3 was Wallace Sinclair of Bound Brook, New Jersey, who was rescued by the *American Shipper*. He had originally been in boat 10, but it was overcrowded so he had transferred to boat 3. He felt he was lucky to survive and was thankful it was over. Sinclair was the representative for three American manufacturing concerns in South America.[20]

Mr. Sinclair, looking sharp-eyed and confident, traveling alone, was able to give precise descriptions of the conditions during the short voyage of the *Vestris* because of his long experience in ocean travel. He gave this account:

I left the ship in boat number 10. Before getting in I noticed that the paint on the boat was dead paint. In one place it had scaled off the bottom of the boat over a patch as large as my hand. That was before the boat had been lowered and scraped against the side of the ship. It wasn't scraped anyway, the paint nearby was loose and flakey.

Why our boat did not tip over when it was launched no one knows. There were fifty to fifty-five persons in it. No one was in charge, and those of us who tried to take charge were very unsuccessful. We drifted away and the ship sank in, I should say, about twenty minutes. My position in the boat left me with my back to the ship, so I did not see it sink.

Our boat was settling rapidly, and there were too many in it. Two other boats came near, and we transferred some of our passengers to them. I transferred to boat 3, and we picked up more passengers. The last one we picked up was among the wreckage over the spot where the ship sank. After that we got away from the wreckage for fear it would damage the boat. We had forty-three in the boat when we were picked up by the *American Shipper*. Water was coming through the seams of the boat, and we had to bail constantly with two pails.

So far as I know there had never been any lifeboat drill on the ship. Our boat had biscuits on board, there might have been some water, and there were eight oars. No man on board was strong enough to unscrew the can containing the flares. It was rusted fast. Luckily one of the crew had a hatchet on his belt, and he knocked the top of the can off.

When nightfall came, we went to light the lanterns which were on the boat. We found two new ones, which were empty, and one old one, which was about half full of oil. Unfortunately, when we went to light the old one, the bottom fell out and we lost the oil. No member of the crew who was in the boat knew where the extra oil should have been stored, and as we later discovered in our search, there was none.

Our flares were dry, but they seemed fragile, and about half of them broke when we tried to light them. That is, the lighting caps came off. Luckily, a member of the crew was able to set them off by lighting them himself down in the bottom of the boat, so we were able to use them.[21]

Among crew members who were aboard lifeboat 3 was Warwick Roberts, the ship's boatswain. He testified in the British Board of Trade inquiry.

He said he was thrown forward from a fishplate into lifeboat 3. There were three stewards and a boy in the boat at this time. They had only one oar left, and the rudder was gone. He saw the *Vestris* sink and picked up three or four people before she went down.

"I was picking them up all the time after she sank," he added. "I was told we had 45 persons on board when we were picked up by the *American Shipper*."[22]

Lifeboat No. 10

Lifeboat 10 was the fourth one picked up by the *American Shipper* at about 7:00 A.M. Tuesday morning. [§] One of those who was rescued was Brazilian Jorge B. doValle, a third-class passenger formerly attached to the Brazilian consulate in New York. He gave his New York address as 140 East Sixteenth Street. When doValle informed the Hoover attorneys that he held a mate's license for Amazon River navigation, it lent a nautical air to his testimony. He said his English was "practical." It was until he was asked to describe the difference between a "stiff" and a "tender" ship. There his English failed him.

He testified that the porthole of his third-class cabin, well forward on the *Vestris*, was "a sieve." It began to leak on Sunday when the list, which he said was "very little" on Saturday, grew worse. The port had no gasket, and the glass was cracked. A steward tried to fix it but failed. He said that the weather on Sunday was awful, "plenty wind, plenty rough."

The storm forced a lot of water into the hallway outside his cabin and covered the floor half an inch deep. He stated, "Every time a wave come, water fall inside."

His cabin mates wakened him early Monday morning, telling him about the list, which he said was "terrible." He dressed but omitted his shoes, which would be a hindrance if he had to swim.

His experience in lifeboat 10 was unpleasant for doValle. He believed it was launched by passengers not crewmen but later admitted he could not know that for certain. He told of the problems in lowering the boat down the port side of the ship, which was the high side.

Since he knew the rule of the sea—women and children first— he waited for the order for the men to take to the boats before getting into the one he chose, lifeboat 10. The boat was crowded, but later a number of passengers transferred to other lifeboats.

None of the crew was aboard lifeboat 10, doValle said. There was no water in the butt, even though the plug was there. The lifeboat leaked so badly that it had to be bailed all night. He said he could neither bail nor row, because his hands were crippled and numb with cold. He occupied his time with "thinking of his fate."

The No. 10 lifeboat was picked up shortly after dawn by the *American Shipper*.[23] Sunrise at the location where the lifeboats were found on November 13, 1928, was 6:20 A.M. EST.[24] †

§Among the most bitter of those rescued from lifeboat 10 was Orrin S. Stevens, who worked in Buenos Aires for the First National Bank of Boston. Mr. Stevens was returning to his post with his bride

---

† Standard time was used exclusively between WW I and WW II. Daylight Savings Time was not extended into November until 2007.

of three months. She did not survive. Overcome with emotion, he broke into a furious condemnation of the officers of the *Vestris*.

It was criminal negligence! The officers were all murderers!

When my wife and I were in the Argentine Consulate on Saturday to get our visas, an attaché there [Carlos Quiros], who said that he was going on the *Vestris*, told us he had been aboard the ship and that it was listing perceptibly at its pier.

The officers failed to use their heads at all—they were brave enough but they had no brains.

The slim, pale banker tapped his forehead to make his point. The phrase evidently soothed his anger somewhat, and he kept repeating it and tapping his head as he continued his story:

The officers kept telling us that there was no danger. They said that the watertight compartments would keep the ship afloat indefinitely. If they had only sent out a radio appeal for help Sunday everyone could have been saved. As it happened, I understand that when the SOS was finally sent out our position was given thirty miles wrong. They failed to allow for our drift and that is why the rescue ships had so much difficulty in locating us.

The position radioed by the *Vestris* was evidently correct; the fault lay with the rescue vessels, which proceeded to the SOS location expecting the debris field to still be there. Of course by then the debris and lifeboats had drifted some thirty or forty miles as they were caught in the middle of the Gulf Stream. Stevens continued with his account:

There was no panic. The passengers were splendid even when they realized the ship was doomed. I didn't hear a single scream even. The officers and crew were brave, too, but they had no brains.

They never told us that the ship was sinking. There was no alarm, no warning to go to the boats. When the list got so bad we

couldn't stay in our staterooms, my wife and I went on deck and
saw that they were launching the boats.

I put my wife in one of the boats, which had a number of
officers in it, thinking that that was the safest place I could find
for her. Puppe did the same with his wife, and then he and I went
along the deck toward another boat.

I didn't see what happened to the boat in which I put my
wife, but I have been told that one of the davits broke and the
boat was torn in two as it was being lowered to the sea.

Overcome by the recollection, he stopped while several passengers
who had been with him in lifeboat 10 did what little they could to
comfort him. One of them said that his fortitude had been an inspi-
ration to them throughout the trying night. He not only kept bailing
the leaky craft unceasingly but kept the others working when they
would have stopped from exhaustion.[25]

§Another on lifeboat 10 was Cline F. Slaughter of Chicago, a traveling
auditor for the International Harvester Company. Happy to have
found that his wife—whom he had given up as drowned—had been
rescued by the French tanker *Myriam*, he was not as severe in his
language. But he agreed that there was much to condemn in the
behavior of the officers and crew of the *Vestris*:

Captain Carey was too slow in sending out his SOS. He also
waited too long to lower his lifeboats. In fact, there was general
inefficiency on the part of the officers and crew. They kept telling
us everything was all right until ten or eleven o'clock Monday
morning, even after the ship was listing more than 30 degrees.

Sunday evening when the big wave hit us which stove in a
supposedly watertight port, my wife and I were asleep and slept
right through until Monday morning. When we woke up the ship
was listing very badly but the crew didn't warn us of any danger.
We made our way up on deck and saw the situation for ourselves.

The crew kept their heads pretty well but they were too slow getting started. They should have had a ship stand by Sunday when the trouble first developed.

My wife and I got into a lifeboat on the port side but the ship was listing so much by then that the boat couldn't be launched. We got out and walked down the sloping side. As we did so the *Vestris* went under and despite our life belts we were carried down by suction.

We had clasped hands as firmly as we could but our grip was broken and within a minute or two it seemed my wife was carried a hundred yards away from me. I saw her hoisted into a lifeboat as I floated away in another direction.

Pretty soon a floating plank came along and I clung to that for a while. Then what seemed like a piece of chicken coop came by and I hung on to that. I was in the water for three or four hours before lifeboat 10 picked me up. I was pretty well bruised from hitting the side of the ship when it went down but I'm too happy now that I know my wife is safe to worry about that.[26]

"The passengers of the ship were courageous throughout the entire ordeal, but this is not true of the crew," asserted Edward John Marvin, a passenger on the *Vestris* who was among those rescued by the *American Shipper*. He was employed by the Standard Oil Company of New Jersey as an auditor and had been on a business trip to Trinidad by way of Barbados. Resting in his home at 17 Brookfield Road in Montclair, New Jersey, thirty-four-year-old Marvin was accompanied by his wife and two children, Joan and Howard—whom he brought a piece of the life belt he wore as a souvenir—as he told of the disaster. He told the interviewer:

No lives would have been lost if the SOS had been sent out at the proper time, and the lifeboats had been in condition and launched many hours before they were. The lifeboat I was in had

twelve flares, but only three of them worked. It was fortunate that
we were picked up, for it was a flare from a lifeboat about a mile
away that signaled the *American Shipper*. The lifeboat leaked
constantly and it was necessary to keep bailing out the water to
keep it afloat. During the night we experienced lightning and
thunder, and it rained and hailed intermittently.

The officers were all right, but the crew didn't know a thing
about manning lifeboats. It was the general opinion that the
Vestris was not in proper condition to leave port. A member of
the crew was heard to have made this statement. If this is so the
blame should be pinned on the captain for leaving port, or the
steamship company for allowing it to leave port.[27]

§At the US Steamboat Inspection Service hearing led by Hoover,
Marvin told of seeing women and children wearing life belts huddled "pitifully" in a corner of the smoking room when he got up
shortly after seven o'clock Monday morning. He stated that this sight
convinced him that the captain had already sent out an SOS.

Early that morning Edward Marvin also told of seeing seawater
outside his porthole. He was in cabin 89 on the dining room deck of
the *Vestris*. He also described the violent roll of the *Vestris* on
Sunday night. He jeered the crew's inept attempts Monday morning
to cast away cargo manually after finding that the winches were
without power. He described the notorious "hole" in No. 8 lifeboat,
and how a passenger had to use a penknife to cut the falls of lifeboat
10 when the boat's hatchet failed to do the job.

When he came on board Saturday, the day the *Vestris* sailed,
Marvin saw nothing unusual about the ship. Sunday morning he
estimated from the forward promenade deck that the wind was
blowing from sixty to eighty miles an hour. The chief officer
confirmed this. Marvin then observed waves breaking over the
vessel on the port side, and stated that there was a slight list.

Sunday afternoon, he saw by leaning over the starboard rail that the ship was not moving. He listened for the sound of the engines but could not hear them. He then knew that the *Vestris* was stopped. It was at dinner that night when the ship violently rolled, and the chief officer "dashed madly out." He said that the scene was chaotic. In the salon lounge [a cocktail bar] all of the chairs and tables had slid into a corner on the starboard side of the room.

Early Monday morning, he saw the water outside his porthole. He stated that the list was so severe that nobody could move on deck without the aid of ropes. When he came into the smoking room, he saw the women and children huddled in the corner. Mr. Marvin watched lifeboats Nos. 4 and 6 as they were loaded shortly afterward with women and children. He described how it was nearly impossible to lower the boats because they caught their clapboard sides on the hull of the *Vestris* and had to be pushed off with oars.

He described the hole in No. 8 lifeboat as big as his two hands held together. It looked to be a new hole, he said, and it must have been broken in before launching since he did not believe it could have hit against the hull when he saw it. He added:

> I want to say that the passengers were the most courageous crowd I ever saw in my life. Not one of the women was crying. But after placing the women in the boats there was a mad scramble of the crew for places for themselves.
>
> How number 10 lifeboat ever cleared the ship I don't know. When we were launched we had great difficulty cutting the ropes. The axe in the boat was too blunt. Finally somebody produced a penknife and hacked us free. I felt sure that lifeboats 4 and 6 were safe because the women and children were in them, but I didn't see them again.[28]

Marvin said that of the twelve flares in the boat, nine were defective. Passengers tried to light all of them during the long night. The boat's lantern had no bottom, so it was useless, and the boat's tiller was broken and unusable. Marvin saw no evidence of cowardice on the part of the officers and crew. His only complaint was that some of the crew in lifeboat 10 refused to bail.[29]

§ First-class passenger Edward J. Walsh of 137 Eighty-eighth Street, Brooklyn, who was in the construction business, declined to discuss the conduct of the crew of the *Vestris*.

"I don't want to say anything about that now," he said.

"That's right!" agreed William Bernard Burke, auditor for the Electric Bond and Share Company. "But we'll say plenty about that when the official inquiry opens."

The *Vestris* ran into rough weather Sunday morning, and about 11:00 a.m. it hove to, Walsh said. He believed that water was leaking through the portholes. He gave this account of subsequent events:

> About three or four Sunday afternoon, I realized that we were in grave danger. The ship's list kept getting worse and worse and about 6:00 p.m. it was so bad that the tables and chairs, which were fastened down, ripped loose.
>
> I sat in my stateroom all night Sunday just waiting for it to happen. The next morning the list was so bad that you couldn't walk on the decks. You had to make your way clinging from one thing to another.
>
> There was no panic among the passengers, not so much as a single scream from one of the women. They can't say that about us anyway. And there was one junior officer I very much admired for the way he tried to get the passengers into the boats.
>
> I got into lifeboat 10, which was badly overcrowded. It had about sixty people in it, about half of them passengers and most of the rest cooks and waiters. The boat slid along the sloping side of the ship and kept hanging up as it was being lowered.
>
> As we reached the water the bow of the boat was dashed against the side of the ship, and it is a wonder that we were not

capsized. We cut the ropes from the davits and managed to get away. After about ten minutes, we had gotten 100 yards from the *Vestris*, when we saw its bow go up. Then it rolled over on one side and slid under.

Our boat was leaking and shipping water badly. Three or four men were at the oars, while the rest of us were busily bailing. We had only one water bucket, one wooden pail and some tin cans, and the boat was leaking so badly that we figured it was bound to go down.

Half an hour later we transferred eight or ten people to another boat which was only about half full. There was quite a sea running and it was a very difficult operation, but it relieved matters in our boat a lot. After that we just drifted in the Gulf Stream, letting it take us where it would.

Everybody got very weary but we had to keep bailing. There was always five or six inches of water in the bottom of the boat in spite of our best efforts. It was raining and at times hailing, and when we got cold we got into the warm water of the Gulf Stream to warm up.

Around nine in the evening we saw the beams of the *American Shipper*'s searchlights against the sky, but so far off that we couldn't see the lights themselves. We kept drifting through the night, perhaps as much as fifty or sixty miles, and occasionally we would see searchlights in the distance.

Not long after 4:00 A.M. the searchlight of the *American Shipper* came into sight once more and this time came steadily toward us until it picked us up and held us. The *Shipper* came alongside us about 7:00 A.M. and the crew hoisted us up ladders.

I want to thank the officers and crew of the *American Shipper* for their wonderful skill in finding us on a pitch dark night and for the way they maneuvered the ship to pick us up and for the splendid treatment they gave us afterward.[30]

Survivors probably in lifeboat 10

Two of the more prominent survivors among the passengers of the ill-fated *Vestris* were Dr. August Groman, of Odebolt, Iowa, and William Phipps Adams, wealthy businessman of Chicago and a rancher in Odebolt. We have some excellent accounts from these two, but neither of them mentions the number of the lifeboat they were in. Most likely it was number 10. William Adams said their lifeboat was launched from the port side at about two o'clock in the afternoon. Only two lifeboats were successfully launched from the port side: lifeboat 10 and lifeboat 14. Dr. Groman stated that the man in charge of their lifeboat was a "big black crewman," but the man in charge of lifeboat 14 was Lionel Licorish, who was described as a small man. Dr. Groman also said that they were in the lifeboat for seventeen hours before they were picked up by the *American Shipper*. Seventeen hours later would make it about seven o'clock the next morning—the time Walsh said lifeboat 10 was picked up by the *Shipper*. A preponderance of the available evidence, circumstantial though it may be, is that they were in lifeboat 10.

Dr. Groman sums up the situation quite well in an interview by John C. Metcalfe, a reporter for the Chicago *Southtown Economist*:

> One by one the boats were lowered. It was less than fifteen minutes before the *Vestris* sank that my friend and I entered one of the boats. It had a crew of colored men, with the exception of five whites. We dressed the boat perfect—that is, we had equal weight on each side—and then proceeded to row away.
>
> The *Vestris* went down a few minutes later and we drifted into the night. We did not know which way to head and therefore drifted a great deal, also reserving our energy. We knew that the *Vestris* had drifted a long way from where she had sent her last SOS and so we tried to steer into her original position.
>
> It must have been about four o'clock Tuesday morning when we sighted the *American Shipper,* one of the relief vessels. But we

could see them only when the waves carried us high into the air. Otherwise we were hidden in the water. It was dark and the *American Shipper* was using her searchlights to find the lifeboats.

We tried to get into the path of the lights, but our efforts were futile. Finally we conceived the idea of yelling in unison like the students do at a college football game. We yelled together at the top of our voices. Again our efforts seemed hopeless, but we kept at it. Finally somebody on board the *American Shipper* heard us and in a minute a searchlight shone near us. We yelled louder, despite our exhausted condition. At last the light found us. It seemed to revive us and we cheered like mad men.

The next thing to do was to get alongside the *American Shipper* without smashing our boat to bits. The ship, however, was fully prepared for us. There were ropes lowered on all sides of her to which we could hold using our oars to keep the lifeboat away from the side of the steamer. Even with this precaution, however, we hit the side of the *Shipper* hard several times when waves overpowered us.

The *Shipper* dropped rope ladders to us and the colored men climbed them [quickly and easily]. The rest of us were too weak. We called for a bowline, which was immediately lowered. And in a minute we were lifted out of the ocean.[31]

In a newspaper interview immediately following his rescue, Dr. Groman had this to say about the events and who was responsible for their survival:

We were in the last lifeboat. The first were women and children. When it was full, they began to lower it as quickly as they could because the ship was listing dangerously. Everyone was dumped into the water—women with babies. It was terrible. Some boats got off, others could not be launched at all. Some foundered with too many people. We saw the officers speeding toward the coast in a motor launch with an awning. The ocean was full of people. It was dark and freezing cold. Those in the water would try to cling to the sides of the lifeboats, but their hands were beaten off with the oars. We had a big black crewman in our boat. He probably saved our lives. He took charge and sent up the two flares we had. We were seventeen hours in that boat before we were picked up by the *American Shipper*. We saw our black crewman there. Mr. Adams reached in his pocket and gave him a roll of bills, all he had with him, maybe two thousand dollars. We would not have survived without him.[32]

At seventy-two, Dr. Groman was the oldest survivor of the *Vestris*. His traveling companion, William Phipps Adams, had this to say about their experience in the lifeboat, written in a letter to Judson Jackson, in answer to a request for information:

The sea was rough—12 to 18 ft waves and a good ground swell running and you will realize that only at certain times could one see much in such a sea way—and some of these things I saw and some I heard of from others when I got on the *American Shipper* the next day.

It was about two o'clock in the [afternoon] when we got away from the ship and as the boats were rocking about and in danger of collision, we made all haste to get away from them and also from the suck of the sea as the ship sank as well as floating debris that would be there soon after. We had 47 in our boat and could take no more—6 or 7 were white—I did most of the "talking" but it didn't do much good. The men were laborers—not sailors—and anyway, it was everyone for himself. It took four to bail our boat all the time. How I did wish for my Colt automatic (in my trunk) and a couple of feet of garden hose to make these Negroes bail when I said and sit down when someone saw a "brilliance" on the horizon and called out "Ship!" He nearly tipped the boat over—we had no officer on our boat—nor did I see any on the ship, but they may have been getting out boats on the starboard side.[33]

Another who was in lifeboat 10 was Fred W. Puppe, a German national. The *New York Times* stated "Mr. Puppe went in lifeboat 10 after he had seen his wife and child placed in lifeboat 8, which was one of the boats that sank, and he did not see his wife or child again."[34] His account consists of the testimony he gave in the Tuttle inquiry.

§ Puppe, the first to testify, explained that he was an electrical engineer, and was traveling to South America with his wife and seven-month-old baby to a job with an electric utility company there:

> Monday morning there was no coffee, tea or drinking water. They were changing passengers about during the morning.
>
> At nine o'clock I took my wife and baby up to the smoking room. I took the straps off our luggage to fasten around them to help them get along. Later I found a place between the wall of the smoking room and the tilted deck where they could hold fast in the "V" which was formed there.
>
> I thought then that no SOS had been sent out. Later I was told that one had been sent out at eleven o'clock, and that a ship would be standing by in twenty-five minutes. There was no panic among the passengers, only a few women and children were crying.
>
> There were no officers anywhere around and throughout the whole time I did not see an officer give an order. I could see that those trying to lower the boats had never had any experience in lowering boats. By 1:30 that afternoon not one was in the water. They did not have the tools they needed, and the equipment was not right. I saw them try to use their hands to force tackle that did not want to go. I saw members of the crew running from boat to boat, but I did not realize then what they were doing.

> Then word was passed for the women and children to get in the lifeboats. I went toward one aft with my wife and baby, but we were ordered forward. They took my wife and the baby on one and I started to follow them. There were other husbands in the boat with their wives and families. But the captain, who was standing on the bridge just above us, shouted down not to let me in.
>
> "Don't let that man on," he called down. "Don't let another man aboard. You go get in another lifeboat. Your wife will be saved hours before you will be."
>
> I got into another lifeboat and it was launched. As we pulled away I noticed that two lifeboats were still fastened to the ship, part way down to the water, and that there were people in them. Then the ship sank. I didn't imagine that my wife and baby were in that boat.

When Puppe reached this point in his story, his voice choked, he clenched his hands, and he teared up. The crowded courtroom fell silent. Mr. Tuttle appeared embarrassed to be there. Commissioner O'Neill sat silent, and the hearing stopped until Puppe regained control of himself.

The witness then related what happened Monday afternoon and night as the lifeboat he was in drifted. They had to bail all night, he said, using five jugs and cans. He estimated that about ten gallons of water were bailed out every minute because the boat leaked so fast, and he said this rate had to be sustained for the sixteen or seventeen hours that elapsed before they were rescued.

> We got out the keg of drinking water stored in the boat and found that it was sea water. We got out the flares and found they were wet and useless. Then I realized that the members of the crew I had seen running from boat to boat had been taking the provisions and good flares to the boats for themselves.

Answering a question from Mr. Tuttle, Puppe said he estimated the list of the ship to have been ten degrees on Sunday, eighteen degrees early Monday morning, and thirty degrees or more later on Monday morning.

> At night we saw there were flares in some of the other boats, and then I realized what it meant when I had seen those men taking things from one boat and putting them into others. They were fixing a few with proper equipment, planning to ride in them themselves, and they winked to their friends to join them.

As Puppe finished his testimony there was a stir in the hearing room as some men, whose hands and faces bore fresh scars, entered and sat down. They were evidently *Vestris* crewmen.[35]

Lifeboat No. 14

Lifeboat 14 was likely the last one rescued by the *American Shipper*, picked up by 7:30 A.M. on Tuesday. One of the more captivating occupants of lifeboat 14 was the man who was in charge of it: Lionel Licorish, a small man from Barbados, who was quartermaster of the *Vestris*. Not only did he keep his own charges safe from the sea, but every now and then he jumped into the waves to rescue some unfortunate swimmer.

Passengers aboard the *American Shipper* put the number of Lionel's rescues at more than twenty, but he himself would not talk about it. Newspapermen found him sitting on the rail of the *American Shipper* looking nonchalant and happily hanging his legs over the side. He smiled but said nothing.

This quartermaster with the quaint name got into a lifeboat containing an injured fireman. There were no oars, so he swam to a capsized boat to get oars. Then they rowed back and forth through the debris for hours until they had picked up all the survivors they could. Sixteen was his count of those he saved; others said it was more like twenty.

"That little crewman did what the officers of the *Vestris* failed to do," affirmed Alfredo Ramos, who was in the water fourteen hours before Licorish pulled him aboard lifeboat No. 14.[36]

§ One of the most vivid accounts was told by Captain Frederik Sorensen, who was rescued from lifeboat No. 14.

Captain Sorensen, who held a master's license and made his previous voyage as chief officer on the *Henry S. Grove*, had been out of the Marine Hospital on Staten Island for only a week before he boarded the *Vestris* as a second-class passenger. He was bound for Barbados to recuperate from his illness. Wednesday he appeared pale and worn out, wearing old trousers and a jersey, leaning against the rail of the rescue ship that carried him up the bay from Quarantine. He was going to his New York residence at the St. George Hotel in Brooklyn.

"I know something about ships myself," Sorensen declared, "and I can say that the reason the *Vestris* foundered is due to criminal neglect. The captain was to blame when he sent out the first SOS and then canceled this message. He should have had other ships standing by long before the passengers had to take to the lifeboats." [He would later deny having made this statement.]

He said that the boiler room was flooded on Sunday night and that the ship listed so badly by Monday morning that he was forced to walk with his hands on the deck for support.

"The bad weather really started Saturday night," he continued, "and the ship [began listing the next day]. We couldn't get any information about the trouble from the captain or crew, and when the engines had stopped and [we] were lying helpless Sunday, the captain said there was no trouble and we were proceeding at once, which was not so."

Captain Sorensen said he left the ship about one o'clock Monday afternoon in lifeboat 8, with around forty or fifty other persons, mostly women and children. The lifeboat soon swamped.

Just about that time I was washed out by the sea. Swimming near me was Hermann Rueckert and a stewardess. We all had life belts on and swam in the water, which was warm as it was in the

Gulf Stream, until about six o'clock in the evening when we were picked up by lifeboat 14. Just when I saw the boat, Rueckert was exhausted [and] about to give up, but I urged him to keep going until I could get to the boat.

I swam to the boat, in which there were about seventeen persons, and we rowed back and picked up the other two swimming in the water. They only had two oars in the boat and a colored quartermaster, who was in charge, steering with an oar as there was no rudder. The purser, Mr. Pugh, who had also been picked up from the water, was lying on the bottom of the boat.

They stayed in the lifeboat until early the next morning, when Sorensen saw the searchlights of the *American Shipper* scanning the water, looking for survivors.

Sorensen was enthusiastic in his praise for Captain Cumings and the work of his crew in rescuing survivors. "It was a great thing they were doing that," he said, "because otherwise we might not have been picked up. They helped us aboard the ship by means of a ladder and hoisted up an old woman who was hurt in a cargo net."[37]

The old woman who was hoisted on board the *Shipper* in a cargo net was Miss Helen Cubbin, a stewardess from the Bronx, New York. She told how she tried to save a little boy and a little girl. Routed out of her cabin by a sudden surge of water, she only had time to pick up the two children before getting onto a lifeboat.

The port side of the *Vestris* was high in the air because of the list to starboard. Our boat had to be worked out over the hull sidewise until it was clear enough to drop. When they lowered it there were only about four men in it. I had the little girl by the hand and the boy snuggled against my shoulder, both of them screaming in terror.

When our boat hit the water it capsized and both children were thrown into the water clear of me. They must have gone down immediately for I did not see them again and I have no idea who they were. There were [also] other children in the boat who were never seen again.

I swam for a while and then was picked up by [lifeboat 14]. But that boat was leaky and we were kept bailing all the time. We spent the whole night on the water with nothing to eat but biscuits. I couldn't eat them. The next morning we were picked up by the *American Shipper*.[38]

§ "It was murder," affirmed Hermann Rueckert, a businessman of Leipzig, Germany, who was rescued from lifeboat 14 by the *American Shipper*. After the list was noticed Sunday afternoon, Mr. Rueckert said he was forced to change his stateroom three times owing to seepage into the rooms. He began his story with this:

The captain sent an SOS call on Sunday night at eleven o'clock and cancelled it soon afterward. There would have been plenty of time to save every one if the SOS had not been cancelled as we learned later. The captain did not call for help again until Monday morning at ten o'clock. Then it was too late to save many. [Ed. Note: The captain did not authorize an SOS on Sunday.]

Mr. Rueckert helped a woman and two children into lifeboat No. 6. Rueckert said it took more than an hour to lower this first lifeboat, loaded with women and children. It sank almost instantly when it finally reached the sea.

He said the crew was not openly unfriendly to the passengers, but they often disobeyed orders from the few officers he saw:

What panic there was, was confined to the crew. The white passengers were calm throughout the disaster. Most of the women on port side of the ship, the left, were in lifeboat number 6, which proved to be the most unseaworthy and sank almost as soon as it was in the water.

On the other side of the ship some of the lifeboats with the crew left with only six to ten in one boat. The captain called after some of these boats which had been taken by the crew, but they paid no heed. Only one boat returned.

He said the overcrowding of his own lifeboat, No. 8, caused it to fill with water despite desperate efforts made by the passengers, most of whom were men, to bail it out.

That caused all of the passengers to be floated out of the lifeboat into the water. However, they soon came back and all were hanging onto the sides of the boat. When I saw that, a young lady, a man, and [I] left the boat and started to swim away to try to make another lifeboat. The first boat we saw did not take us up, probably because they were overloaded. Other boats, manned by the crew, rowed away from us. I saw no one actually repelled by the crew, but I did see the crew row away from seventy or eighty people who were swimming near me. . . .

We swam to boat 14, which picked up the other man and then me. The young lady in the meantime had disappeared. After I was taken into the boat I was utterly exhausted as I had been swimming in a rough sea for over an hour and a half. It was about 7:00 P.M. when this boat picked me up. There were, of course, no provisions or anything to drink in the lifeboats and in fact we had nothing to eat or drink from Sunday night up to Tuesday around seven o'clock when we were picked up.[39]

Walter Spitz of Berlin was a businessman on a pleasure trip to South America. He told of having had to jump into the sea after the lifeboat that he had boarded could not be lowered because nobody could find a knife to cut the ropes holding it. He was picked up almost right away by lifeboat 14.

Rueckert and Spitz said that they had seen the captain on deck just before the ship sank and had not seem him jump off, although they were less than a hundred yards from the *Vestris* when it went down. They believed he went down with the ship.

Rueckert and Spitz were staying at the Fifth Avenue Hotel. Mr. Spitz was the temporary guest of Eugene Miller, manager of the hotel, until he could wire to Germany for funds. Mr. Rueckert had a few dollars in his pockets and was the guest of Henry Tilford who lived at the hotel.[40]

Alfredo Ramos, an Argentine student who was returning home on the *Vestris* after a vacation in New York and Europe, told of the consternation of passengers loaded into lifeboats that could not be lowered to the water. He was in one of them himself, swung out over the waves on the davits and then left hanging there. He and some of the others jumped into the water trusting their life belts to save them. He never saw the others again and had no idea whether they had been saved or had been lost.[41]

He floated around for some two hours before he was picked up by lifeboat 14. When rescued he took off his life preserver calmly. His chief concern seemed to be that the coloring from his bright red tie had stained his shirt.[42]

Mormon missionary David H. Huish continued his narrative:

> I was still on the side of the ship when it went down and a big wave came along with the suction of the ship, and took me under. While in the water I caught hold of a panel, the bottom of a lifeboat or something, and came up. . . . I floated on a 2×12 plank for a few minutes and pretty soon a lifeboat, No. 14, came somewhat close and I swam over to it and was helped in the boat. We picked up about twelve more people—19 in all. I looked constantly in the water and into the other lifeboats for Elder Burt, but he was not to be seen.
>
> The waves soon drove us from the wreck. We had no rudder and only three oars, so we were helpless in picking up any more people. The water was warm at first, but towards evening it became cold, and I never shook so much through fear of facing the public as I did during that night from cold. We drifted all night, without any flares or torpedoes. Two or three storms came up, one hail storm. By morning the waves were very high and we did not know at any time whether we would be swallowed up or not.

About 11:00 P.M. at night came our first hope, when we saw a flashlight. The ship came nearer and about 4:00 A.M. they picked up one lifeboat. From then until 8:00 A.M., we drifted and were finally picked up by the *American Shipper*—the last of the five boats that this ship picked up. We had been trying all the time to get their attention with a flashlight and with our shouts, but to no avail until after dawn.

They gave us something to eat on the ship, and a place to get warm and to dry our clothes. They searched for more survivors as long as there was any hope, and then set sail for New York, and arrived here about 9:00 A.M. We were treated well on the ship, but slept on blankets on the hard floor and were glad to get that.[43]

David Huish told a reporter from the *New York Times*, "I thought I saw Burt swim to another boat, and I was amazed when they told me he was missing. He was a good swimmer."[44]

Rescued from the sea

There were two persons who were taken on board the *American Shipper* directly from the ocean rather than from a lifeboat. These were Paul Dana, the South American representative of RCA (the Radio Corporation of America) who was on a business trip to Buenos Aires, and Clara Ball, a stewardess on the *Vestris*. When the *American Shipper* picked them up, Captain Cumings radioed to shore that they were "two of the pluckiest people I ever met."[45] Here are their stories, intertwined as they inevitably were.

Paul Dana picked up with Monday morning events, starting from where he left off in Chapter Five, page 48:

§ The sky had cleared during the night, and now the sun was shining brightly. The wind still was blowing hard, however, and those waves—great towering black waves—they looked forty feet high. "Well, anyway, we've got a nice day to be wrecked on," one of the men remarked with a grin.

Officers were going about, cheery but non-committal. They never told us anything. I never knew until after the *Vestris* had gone down that they had even sent out an SOS, although when I went into the smoking room for a little while that morning and tried to read and divert my mind, I thought I heard the radio sort of sputtering.

Once I went back to my cabin to try to rescue my money that I had left in a trunk. No use—that trunk was under water and wedged in under my bunk. I got my passport, stuck a pack of cigarettes in my pocket, and left—for good.

The ship kept on tipping. It looked to me as though she had a list of forty-five degrees. You could hardly walk on the deck. At 10:30 the women and children were brought up and told to put on life preservers. Still we got no definite information from the officers or crew.

The women were wonderful. Some of them were crying quietly, but there was no hysteria. One of them held in her arms a baby not more than eight months old. All the children had been bundled up. They knew now that they would probably have to leave the ship.

"Isn't this fun!" I said to one little fellow with bright blue eyes, trying to cheer him up. He looked at his mother, who was crying, and nodded—solemnly.

At 11:30 they started to launch the lifeboats, over the port side—the side that was up as she listed. Then the trouble began. It took them two hours to launch those boats, an operation that usually takes about ten minutes. The difficulty was that the lifeboats kept catching on the sides of the ship. The outsides of the lifeboats were covered with clapboards, like a frame house, and these would catch on the over-lapping steel plates of the ship's hull. They tried to hold the lifeboats out—away from the ship—by pushing with oars; it was a slow and painful process.

Boat number 8—the lifeboat to which I was assigned—had a hole torn in her side as they were letting her down. They patched it up with a piece of tin before they let us in, but the tin didn't hold. When the boats were about twenty feet from the water, they told us to climb down the rope ladder and get in. Two boats were filled with women and children before they filled number 8. In number 8, before I got in, were ten women and children—two youngsters about six or eight years old. One of the [women] was Clara Ball, with whom I subsequently floated on the piece of wreckage. Another was Mrs. Inouye, wife of the counselor at the Japanese embassy at Buenos Aires. The ship's bartender was aboard, and our crew consisted of four Negro sailors.

Three minutes after we pulled away, the *Vestris* keeled over and sank. I think one of the [strangest] things I ever saw was forty men—apparently all those left on board—racing madly down her side and diving off from her keel.

One boat never got away at all. It was still on deck when the ship sank. They had tried to launch five [lifeboats] from the starboard side, but they had to cut them away before they could be filled. These were rowed around to where her keel had been when she went down, and they picked up a lot of the swimmers.

She went down silently—with just a little puff of steam. Her boilers must have been cold. The engine room had not been functioning all morning.

Just a few minutes after we pulled away, our lifeboat began to fill with water. The tin patch tore away. She had air compartments, however, so that we could stay in her even after she was under water. Still the women tried to keep calm, although the tears streamed down their cheeks. There was no screaming. One of the children started to whimper. His mother—hastily drying her eyes—tried to comfort him.

It wasn't long before a big wave came along and capsized us. I got caught under the lifeboat as she turned over. I wasn't hurt, except for a little scrape on my neck. As I struggled out from under, I saw a woman's foot bob up. I grabbed it and hauled her with me. It was Mrs. Ball.

One of the women had drowned when the boat tipped over. Most of the occupants came up, however, including the children. We all got on one side, pulled, and managed to right it again. The women and children climbed in. But the waves were pounding it to pieces. The air compartments, stored under the seats, tore loose. She was breaking up. Great black waves—they looked to be 100 feet high from where we were—kept pounding down on us.

The children disappeared and the women—all except Mrs. Ball. I saw a piece of wooden wreckage floating in the water nearby. "Come on," I said. We swam to it.

"Take hold of one end," I told her, "and I'll take the other." Silently, she did so—never a whimper out of her. We then proceeded to make ourselves as comfortable as we could—to wait for help.

The water was warm—warmer than you find it in an indoor swimming tank. It was so warm, in fact, that you hardly gasped when you plunged in. The fact that it was warm probably helped to save our lives. But it was enervating, too.

Mrs. Ball was fully dressed—with a coat and sweater on under her life preserver. She had on gloves, which certainly saved her hands. Mine are all cut up.

"What in the world is that you have stuck in the front of your life preserver?" I asked her. She grinned and pulled up a slipper. Her feet were so narrow, she explained, that she had a hard time getting shoes to fit, and she wanted to save this pair if she could.

Through the afternoon we clung to the spar. It wasn't so bad, except for the waves. Our life preservers held us up, and having something to cling to made it easier to keep our heads out of the water. Only every now and then a great big wave would come crashing down on us, almost knocking the life out of our bodies and smothering us in foam. I think the hardest thing was to keep from getting excited and floundering about. We had to save our strength and not move about unnecessarily. Mrs. Ball was wonderful. She showed iron nerve. Not a whimper out of her.

Several times during the afternoon we tried to hail the life-boats, but they couldn't hear us or see us because of the waves. I don't think they passed us deliberately. It's pretty hard to spot a swimmer in waves like those.

Toward sunset it began to cloud up. Then a thought flashed into my mind that made me feel a little sick all over. Sharks.

I don't know whether Mrs. Ball ever thought of them or not. If she did, she kept still about it.

During that long night of waves and wind and rain, we clung to the spar—talking intermittently. She told me that she was a stewardess. When we made for the spar and for several hours after we had attached ourselves to it, I had no idea who she was. I confess we spent a good deal of time cussing out Captain Carey for not getting those radio messages out earlier. Mrs. Ball had heard that one went out at eight o'clock Monday morning and had been cancelled. We agreed, though, that it was a tough break for the captain. Nobody likes to admit he's licked until he has to. Probably he thought the *Vestris* would ride it out. I'd like to know if it's true, though, that he sent out an SOS at eight o'clock and then cancelled it. [Ed. Note: It was not true.]

The night wore on. Several times we saw the searchlights of ships. They were too far away to see us then, but it gave us hope. And several times fish brushed against my body in the water and I squirmed; Mrs. Ball never said a word, if she noticed them. If she was undergoing what I was, she never let on at all. I decided to hold on anyway and wait until a shark actually attacked me before I let go.

At daybreak came the worst moment of all for us. As the dawn broke, not a ship was in sight. And our piece of wreckage was beginning to break up. We had to hold it together while using it for a support. At about eleven o'clock we sighted the *Wyoming*. She must have been twenty miles away. But our spar was practically gone.

"How about it?" I asked Mrs. Ball. "Shall we swim for it?"

Without a word she started off toward the *Wyoming*. I kept by her side. We had been swimming for an hour, when I looked back and saw the *American Shipper*. She was only about a mile away, and coming toward us.

I tore off the tail of my shirt and waved it wildly in the air. They saw it. A boat was lowered. As we got up alongside, they threw out a rope. I fastened it about my body under my arms, and grabbed Mrs. Ball. I had an idea that when she could give in, she'd go quickly. I was right.

They hauled us into the boat and then hoisted us up on one of those rope cages they use for hoisting cargo. We never could have made it climbing a ladder. We were much too weak.

I was never treated more wonderfully in my life than I was on the *American Shipper*. The doctor ripped my clothes off, gave me a couple of shots of whiskey and an alcoholic rub, and told me to go to sleep, which I certainly did. When I woke up the steward was ready to shave me. My clothes were all dried.[46]

The following exceptionally well-written essay about the saga of Clara Ball and Paul Dana appeared in the *Rockford* (Illinois) *Morning Star*. It gives a human perspective into their story:

§ Out of the backwash of horror that has swept in from the unmarked sea grave of the SS *Vestris*, there remains one episode that is unique as a study in human values. Already the names of Clara Ball and Paul Dana have become linked in a glowing tale of courage and an epic of survival. Already they share what glory may lie in the curt comment of an old seaman—"the pluckiest pair I have ever met."

Yet had Clara Ball met Paul Dana a few weeks ago she would have given a polite "Yes sir" or "No sir" to some request for service. She would have brought clean towels to the stateroom or supplied a fresh pitcher of water. None ever would have so much as suspected, or had reason to question her potentialities in a grim and terrible crisis.

For Clara Ball was just another stewardess aboard a ship—a role involving more than average anonymity, since a boat is a very tiny world, and once it cuts loose from its port only a fraction of this tiny world is passingly aware of its servants.

And Paul Dana was a wealthy corporation representative from South America.

But when they did meet, Clara Ball was clinging to one end of a bit of wreckage and Paul Dana was clinging to another. Thus, with waves breaking over them from time to time and threatening to hammer loose their stubborn hold on a bit of spar, they faced death together for [more than twenty-one] hours.[†]

What do people talk about and think about under such circumstances one is inclined to wonder.

It was some three hours after they had been tossing about helplessly, seemingly headed for certain death, that it occurred to Dana to wonder about the identity of his partner in a violent journey to eternity.

"Who are you?" Dana asked finally, his voice carrying weakly through the splash and churn of the sea. Clara Ball, ill from swallowing pints of sea water, actually smiled at the question. Surely this was a strange place for introductions and the

---

[†] The article had "twenty-three hours," which is impossible. The *Vestris* sank at 2:36 P.M. on Monday and they were picked up about noon on Tuesday.

observation of social conventions. So she answered: "Oh, I'm just a stewardess from the ship."

It suddenly occurred to both of them that identifications are man-made trifles; tags and brands to mark the offspring of a Smith from the offspring of a Brown. Bobbing in troughs of tropically warm water something happened to the social philosophies of a stewardess and a wealthy corporate agent.

At such an hour, with a long rain-swept night ahead and weakness creeping on with the rising sea swells, one is grateful for courageous and strong company—Dana and Clara Ball will tell you that. It is, perhaps, what kept these two strong and courageous persons going until help arrived.

Did they speak to each other of death?

Not once! Death became a tabooed word. Both thought of it—many times. But to speak of it was to encourage despair.

Nor did one communicate to the other anything of the forebodings that came in the long night hours. Both felt fish brush by and thought of shark-infested waters. But when they spoke it was to say, "How are you coming?" or "Great work, just hang on!"

Did life seem terribly important then?

Clara Ball smiles quietly today when the question is asked.

"It's funny that you don't think of it in terms of life," she says—or words to that effect. "It's just the idea of hanging on somehow. The idea of sticking it out as long as you can. I don't believe you wonder whether life is sweet or bitter. I can't remember wondering much. It's just something that you do instinctively, I guess. Just hanging on with grim despair."

And so Clara Ball and Paul Dana are alive today.

Insofar as anyone knows at this moment, Clara Ball may be a stewardess again tomorrow. And Paul Dana will go back to South America and be an agent for a big corporation.

But the world will know them for years to come as "the pluckiest pair"—and their names, once so unimportant, will be inseparably linked in a saga of the sea.[47]

§ Clara Ball, 38, of Pleasantville, New York, was on her first voyage as a stewardess. She was not very talkative and generally declined to give interviews with the press. The *New York Times* said that Mrs. Ball, agitated and worn out but not showing this until the rescue ship passed Staten Island and all danger was behind her, could not speak at first. Her face quivered nervously, and she wrung her hands. Dark circles showed under her eyes. She said:

> I have always been used to the water, and all I can say is that I found myself in a lifeboat and going down the side. The lifeboat capsized as soon as she hit the water and we were all thrown out. I do not remember how many were in the boat. I found myself in the water and I struck out. I am a good swimmer.
>
> I managed to get to one of the air-tight compartments in the bow of the lifeboat. The boat was floating keel up. The waves were strong, however, and this was wrenched from my grasp. I was swept by waves under some wreckage and was caught head down when I was pulled up. Someone caught my foot. It was Mr. Dana.

The rescue ship's doctor intervened here and stopped Mrs. Ball from continuing. She leaned back in her chair and closed her eyes, obviously exhausted.

A bit later on, Mrs. Ball, her strength evidently returned, picked up her story:

> At first I almost lost hope as we drifted about in the water. Things looked pretty black, but when we saw the lights of the ships that had come to the aid of the *Vestris* I put away all worry and was certain that we would be picked up. I think that I swam for two hours, supported only by the life belt, just before the *American Shipper* came up.

Illustration from *Rockford Morning Star* article

Clockwise from upper left: Clara Ball; artist's conception of Mrs. Ball acting as a stewardess; Paul Dana; artist's conception of Clara Ball and Paul Dana clinging to the broken spar together in the open sea.

I once tried to swim to some wreckage that looked as if it might be more secure than the grating, but the current was too strong for me and I had to go back. It was hard at times to keep from losing my breath. This was during the night when the rain was coming down furiously and the waves were higher.

Mrs. Ball could not be persuaded to discuss her own adventure in more detail. She finished her story with a tribute to the passengers and to the crew:

The passengers showed no panic, and the behavior of the seamen was splendid. They worked like Trojans. There were two Negroes in my boat and they were as calm as anyone. I don't know whether they were saved or not. I hope they were, for their conduct was perfectly splendid.[48]

§After their horrific experience, when they were alongside the *American Shipper*, Clara Ball insisted that Dana be hoisted to the decks first. When they were safely aboard, she refused to be assisted to a chair, declaring that she was "well enough to stand up."

Mrs. Ball walked to a cabin on the *Shipper*. As the doctor bent over her she smiled feebly at the group standing about. Someone asked her if she had been frightened.

Through lips blue from having been cold but in a forceful voice she denied that she had felt fear as she and her companion struggled in the raging sea. "Well," she said slowly, shaking her head, "you can't be afraid when you have got to keep your head."[49]

A passenger, whose name was not given but who was picked up by the *American Shipper* prior to Paul Dana and Clara Ball, made these statements about the pair:

When we got there, Mr. Dana was very weak. We threw two life belts down to them, but Mr. Dana seemed unable to fix his. Then we came nearer and Mrs. Ball was afraid that we might run into them, or that they might be sucked under the boat's wash.

The captain of the *Shipper* threw down a rope and Mrs. Ball left her preserver to swim over and tie the rope to Mr. Dana.

While the man was being pulled aboard the ship, a basket was lowered for Mrs. Ball. It was one of the ordinary mesh baskets used for lifting mail and packages into the hold. Mrs. Ball scrambled into it and was brought up to the deck. Mr. Dana collapsed when he came on board, but Mrs. Ball said:

"I'm all right. Let me walk."[50]

Paul Dana and Clara Ball never saw each other again. Clara Ball went back to working on ships. The crew list for a voyage of the United States Line's SS *President Harding* sailing from New York City July 3, 1931, lists "Ball, Clara G. / born in New York / age 42, height 5 feet 6 inches, brown hair, brown eyes / Hostess."

Paul Dana evidently continued working for RCA in South America. A passenger list for the SS *Western World*, departing Rio de Janeiro, Brazil, August 31, 1933, and arriving at New York City September 13, 1933, lists "Dana, Paul A., age 40, married, born in Evanston, Illinois, November 24, 1892, staying at the Hotel New Yorker in New York City."

∿ ∿

Rescues by the *Myriam*

The *Myriam*, a French tanker commanded by Captain Fernandez Forey, which was bound for Thames Haven, England, was diverted to help with the search for survivors. She rescued two lifeboats, numbers 7 and 11. Captain Forey reported:

> At 2:00 [A.M.] 13th November I saw a red light on the horizon, and on going to the spot found a lifeboat with passengers. We picked them up, and afterwards we saw another red light, [which] turned out to be a flare from another lifeboat, and we took the people on board from her also. From both boats there were 57 people. We cruised around for some hours but could see no trace of any others except for a man, who proved to be dead, in a life belt.[51]

Lifeboat No. 11

Lifeboat 11 was picked up by the *Myriam*, about 4:30 A.M. on Tuesday. On board that lifeboat were Wilma Slaughter and boxer Harry Fay, among others. Fay evidently did not grant any interviews, but Wilma Slaughter—whose twenty-first birthday came the day after she was rescued—told her story.

Wilma "Billie" Slaughter, a pretty, auburn-haired young woman, was the wife of Cline Slaughter. Wilma and Cline were en route to Buenos Aires, where Cline had a position with the International Harvester Company. We patch together two statements by Wilma, one from the *Rockford* [IL] *Daily Republic*, the other from the *Dallas Morning News*:

> When I jumped into the ocean, I was wearing a new gown, a hat, and slippers. When I came to the surface, after what seemed hours, my hat and slippers were gone. And when I was picked up by the *Myriam*, my gown had been transformed into a perfect French creation by shrinkage. I was put to bed in the captain's cabin and they gave me woolen pajamas and warm clothing and then the trousers and blouse to replace my water-soaked clothes. Even now the feminine clothes I have on are not my own.[52]

§I remember getting into a lifeboat with women, children, and members of the crew; I think it was about 1:45 or ten minutes of two. It may have been half an hour later, but the crew could not budge the lifeboat as it was stuck. They ran for a hatchet, then for anything.

Finally, women, children, and members of the crew left the boat I was in, but I stayed and my husband came running with life belts for both of us. I felt that we could get the boat away and that someone would come and help us do it, so I stuck to the boat while other boats were being lowered.

A third-class passenger, Walter Cadogan, allegedly slashed the bow falls of the lifeboat with a razor, dumping the occupants into the sea.

The next thing I knew I was attempting to paddle feebly in the water and felt as though I had come back to life, or was being born again or something. My husband was some yards away attempting to swim against a strong swell, but the sea was too much for him and he could not get near me.

I was almost gone from exhaustion and the [submersion] of more than five minutes—I am convinced it was at least that long before I came to the surface for I must have gone down twenty feet, perhaps more.[53]

Leslie Watson, second officer of the *Vestris*, looking from his lifeboat saw Wilma Slaughter in the water. He jumped into the sea and swam toward her. She said it probably took Watson twenty minutes to reach her.[52]

Finally, after I was floating about for about 20 minutes, trying to get my breath and get the water out of my system, the second officer reached me and swam with me some yards toward another lifeboat, which was filled principally with twenty-seven members of the crew. I must say the crew were good oarsmen, although I cannot explain the fact that there were no women or children in this boat, and I know nothing of how the crew got the boat. I do know they picked me up together with the second officer.

I lay in the bottom of the boat, scarcely able to move and semi-conscious most of the time, and it was only the occasional shrill whistle of men blowing on their fingers to attract attention as they floated in the water and the frenzied screams of women that aroused me from my state of exhaustion.[53]

The screams Wilma Slaughter heard were from women who were in the sea around them, but the lifeboat did not stop to pick up anyone else. As the crewmen rowed the lifeboat, she said she heard some of them say they saw dead bodies floating in the water.[52]

When they were picked up by the French tanker *Myriam*, Mrs. Slaughter did not know that her husband had also been rescued. She commented about the situation on the rescue vessel:

After we were picked up, I can't say what happened for a considerable time because I was so exhausted. I woke in the captain's bed and was treated wonderfully by all of the officers and crew. No American officer or crew could have been more wonderful.

While I suffered considerably from nervous and physical exhaustion when the ship sank, I think I have suffered more from the fact that our rescue ship was unable to pick up a woman and baby we passed early last evening and whom we were unable to locate when the ship turned about to search for them. There was also the pleading eyes of a man wearing tortoiseshell spectacles, whom we passed yesterday afternoon, but were unable to locate after we had put about to search for him."[53]

§Although some rescued passengers accused Captain Carey of being negligent in his duties, Wilma Slaughter did not agree with them. "So far as I know," she said, "there was no such thing as criminal negligence. The captain did his duty and did his best."

As to the accusations made by some that crewmen had rushed into the lifeboats, Wilma Slaughter demurred:

I wouldn't blame the crew. After all, I was in a pretty excited state just before I went down with the ship and don't know what they were doing, except they could not find axes or hatchets to cut loose the boat I was in, which was stuck.[52]

Wilma Slaughter gave this slightly different account of things to another interviewer from the *New York Times*:

§ When she was interviewed, she was dressed to go shopping, a trip delayed by the reporters and photographers. Even in her borrowed clothes, she presented a striking picture: slender and tall, with Titian hair and gray eyes. There was no hint of what she had been through except for some rather nasty red scratches that scarred her hands.

She did not want to talk about her narrow escape, but said:.

Everybody talks to me about it, and I want to forget it. What can I say, except to express my deep gratitude to Officer Watson for twice saving my life. He jumped from a lifeboat and saved me from drowning, and then, as we clung to a door, floating, he saved me again when I lost hold.

Also I cannot say too much in praise of the Captain and crew of the *Myriam*. I learned a lot of French while on board. All day long it was '*merci, merci*.' I can never thank them enough for the wonderful treatment I received.

Asked what she thought about as she clung to the floating door, she replied after a moment:

Well, I remember looking at the wonderful blue ocean rolling about me and the thought flashed through my mind that I would like to have a new dress of that color.

It was terribly hard swimming amid the wreckage, with barrels and chairs and chicken coops bumping into you. It was probably twenty minutes before Officer Watson jumped from a lifeboat into the sea and swam toward me, and we clung to the door for perhaps an hour, the officer supporting and consoling me, until we were picked up by a boat having twenty-seven survivors, including twenty-three members of the crew.

The Negro in charge was very considerate of me, and I was wrapped up in a big coat. I lay exhausted in the bottom of the lifeboat. I could hear cries of despair and whistling from the water, but the boat did not stop to pick up any others.[54]

Wilma Slaughter's husband, Cline, said this in an interview:

My wife and I got into a lifeboat on the port side but the ship was listing so much by then that the boat couldn't be launched. We got out and walked down the sloping side. As we did so the *Vestris* went under and despite our life belts we were carried down by suction.

We had clasped hands as firmly as we could but our grip was broken and within a minute or two it seemed my wife was carried a hundred yards away from me. I saw her hoisted into a lifeboat as I floated away in another direction.[55]

Second Officer Leslie Watson testified at the British Board of Trade inquiry: "I was swimming about in the water and found a lady passenger [Wilma Slaughter]; we got hold of a piece of wreckage, and eventually we were picked up by number 11 boat. . . . We were picked up by the *Myriam* at about 4:30 on Tuesday morning."[56]

Leading Fireman Samuel Augustus Parfitt of Prospect St. James, Barbados, was in charge of lifeboat 11 by "right of seniority" and Second Officer Watson may have been so tired that he did not challenge Parfitt's right to be in charge.

Lifeboat number 7

Lifeboat 7 was the other boat whose occupants were picked up by the *Myriam*. Number 7 was rescued before 5:40 A.M. on Tuesday. Those on this lifeboat were evidently all members of the crew, and few of their stories were published. It seems arguable from interviews with two crewmen, one a member of the ship's band and the other the chief fireman, that they and two other crew members were probably the only occupants of lifeboat No. 7.

Conrad Werner, a member of the band on the *Vestris*, declared that Captain Carey had seemed confused when giving orders to man the lifeboats. He stated that there had been no lifeboat drills on the ship and that the usual notices in the crew's quarters, giving their lifeboat stations, were missing.

When the order to man the boats was issued, he said, they had great trouble unlashing the tarpaulin coverings of the boats. The captain ordered the crew to cut away the lashings, but no one had a knife. There was a delay while the men tried to find knives. When this delay grew too long, Carey rescinded his order.

Werner said that he was thrown off the ship when she lurched from a big wave. After he was in the water and began to swim, he saw the *Vestris* go down. He was picked up after twenty hours in a lifeboat, which had saved him after he had been swimming for an hour. He said that he and three other crewmen, the only occupants of the boat, had been forced to bail continuously to keep their lifeboat from sinking.[57]

Gilbert Ford, chief fireman, who had been on the *Vestris* almost two years, testified at the Tuttle hearings. He stated that he had originally been assigned to lifeboat 9, but when he went on deck he was put in lifeboat 7 on the starboard side. No. 7 was launched without any problems just after two o'clock. There was no officer in charge. The lifeboat leaked at the seams so badly, he said, that they had to bail all night with a bucket and three small pots. The only equipment that he recalled missing was flares. They passed the night bailing steadily before being picked up Tuesday by the *Myriam*.[58]

≈ ≈

Rescues by the *Berlin*

The *Berlin* was a liner of the North German Lloyd line under the command of Captain Eric von Thülen. The *Berlin* picked up the occupants of lifeboat number 13, one man who was clinging to a piece of wreckage, and another who survived with only his life belt to keep him afloat.

Lifeboat No. 13

One of those rescued from lifeboat 13 was Thomas E. Mack, an electrical construction superintendent in Brazil, whose home was in Teckla, Wyoming. He gave this account of that black Monday:

> [S] The *Vestris* was battered by such a storm soon after she left New York Saturday that she soon began to list, and Sunday the waves washed the furniture in the dining room overboard. All that night the storm continued to rage.
>
> In the morning the sea was calmer, but the ship continued listing more and more until by ten o'clock the starboard rail of the promenade deck was under water. Then the captain ordered everybody up and the boats lowered away on the port side. It took two hours to lower four boats to within 10 or 15 feet of the water.
>
> The boats lowered were Nos. 4, 6, 8, and 10. No. 8 had its side stove in. Attempts were made to patch it up, and while the work was going on it was filled with passengers and lowered away.
>
> Women and children were lowered first and all the passengers were very calm. The first mate [Frank Johnson] and second steward [Alfred Duncan] showed the greatest bravery.
>
> The captain seemed calm enough but very undecided as to what procedure to take. Boats number 4 and 6 capsized while being lowered. The ship was lying completely on its side by this time. Only two boats [from the port side] with about 80 people in them were successfully lowered, No. 10 and the stove-in No. 8.

All the rest of the passengers and crew jumped from the side of the ship, everyone swimming as hard as possible to get away from the side of the ship before she finally went down, two minutes later. I swam only 50 feet before the ship went down.

Many passengers were swimming with life belts on all around me, making for the boats. Those that were launched safely were greatly overcrowded. I swam to one which had too many people in it. I turned away so as not to increase the risk. Mr. Maxey and I swam around until we located each other. Then we swam for about two and a half hours until we located boat number 13, which had broken loose from the *Vestris* and had been caught by some of the crew in the water.

Mack said that about eight o'clock Monday night they saw three other lifeboats being rowed, two of them with sails up. Later he found that other passengers who leapt from the *Vestris* had found lifeboats that floated away when the ship foundered.

The six boats[†] stayed close together until a sleet storm separated them about nine o'clock. Later, the people in lifeboat 13 saw search-lights and rowed toward them. At two o'clock in the morning, he said, his lifeboat reached the ship with the searchlight and a tanker lying nearby.

They rowed around the tanker and shouted to no avail. They had to ride out the night in their lifeboat. The sea was very rough and even if they could attract the attention of a rescue ship, they had serious doubts as to whether they could be taken aboard.

At dawn they sighted the battleship *Wyoming* and several other ships, including the *Berlin*. The *Berlin* swung alongside in what Mack said was "a masterly fashion" and took them aboard.

[†] The numbers of these six lifeboats are unknown. The witness did not identify them. We know only that Mr. Mack was in No. 13.

Mack said he was sure that only two boats had been successfully launched from the port side, Nos. 8 and 10, and he believed that No. 8, the one with the hole in her, must have sunk almost immediately.

He described the scene where the *Vestris* sank as was one of horror. Dead bodies held up only by life belts floated among the living, and the few lifeboats were awfully overcrowded. They passed one boat, he said, where the people were sitting in water up to their waists.[59]

Another who was on board lifeboat 13 was Carlos Quiros, Chancellor of the Argentine Embassy in New York City, who was going to Buenos Aires to visit his mother. He supplied this account in writing:

§ [Monday morning] I dressed and went on deck. Other passengers were out there holding onto the railing. The crew was at work at the pumps and they were throwing over some of the lighter pieces of cargo. The second steward, a man named Duncan, came along. He was a very brave man as it turned out. "Listen," I said, "I'm not a captain or anything like that, but I'm pretty sure this ship won't survive another roll." He told me not to worry. Two or three women also appealed to him but he assured them there was no danger. If he knew it he did not show it.

In spite of all this, however, I was by no means so sure. I went to my stateroom again and got my watch and some money. Then, taking a life belt, I went out on deck once more. Men, both passengers and members of the crew, told me life preservers were unnecessary, but I held onto it.

After that there were two or three big rolls and everyone came out on deck. The vessel was now almost on her beam ends and about eleven o'clock the order was given to get off the boats.

Now I come to the most painful part of my story. Although at first there seemed to be no disorder, there was not shown a practical way of boat work. For two hours men struggled with the boats. One boat had a hole in it and the second steward and another man brought a hammer and some nails and patched it up. Women and children and some of the crew got in, but it filled and sank and I saw them next in the water.

Suddenly above the noise and confusion of it all I heard a voice above the others. It came from the bridge. I looked and saw Captain Carey in an overcoat, but with no hat, shouting orders. There was a worried look on his face. His words were shouts and not commands. Still he was calm enough, directing women to go to this boat or to that. There were others there on the bridge with him. English officers and some Negroes. They were trying to save the ship, although they could not do it.

Again my eyes went to the boats. I heard a voice near me shout: "And you like to call yourself a British subject? Go to your post!" A white officer was addressing a Negro member of the crew. He pulled his revolver, but the Negro snatched it from his hand and threw it into the sea. Turning away I saw two boats in the water and the only occupants were members of the crew. I don't know how they got preferred places. Perhaps they had booking agents.

With me through most of this was my old friend William W. Davies, the New York correspondent of *La Nación*. "Stay with me," I told him. "I'm going to live." And when he asked me why, I replied: "Because of my mother." At this time Duncan, the steward, was giving out fig cakes as if it were tea time. An English stewardess was smiling and cheering up the women.

At last Davies and I were alone on the ship with the exception of the men on the bridge. A boat was trying to get off. I went to it and started to get in. In fact I did put one foot in but I pulled it out. The boat was overloaded and half filled with water. I knew she would capsize. "I'm not going to get into something that I'll have to get out of," I said, and they pushed off. I turned to find

Davies, but he was not in sight. No one was in sight there on the ship except those men up on the bridge. There was no use of my staying anymore. I jumped into the sea.

I do not know how long I was in the water. I remember looking at my watch just before I jumped. It was then 2:34. Two minutes later the *Vestris* sank. There must have been sixty persons in the water but the one I noticed was Duncan, the cake-serving steward. He was shouting orders like a traffic cop, telling this man to swim for that boat or that woman to seize some nearby wreckage.

I swam to boat number 13. At that time there were in it only six Negroes. They took me in. There was water in the boat and I began to bail it out. We were afraid the wreckage would stave in our boat and we tried to avoid it. Then we saw a man and picked him up. It was Duncan. Altogether we picked up fifteen men, one of them was the Chief Engineer, named Adams.

The cries of the drowning were pitiful. At last it became dark. Two boats joined us and we tried to keep together. We shouted for help. Each little voice seemed so utterly futile that we shouted in unison. Someone would cry "One, two, three," and then with all the power in our lungs, we would shout "Help!"

But help did not seem to come. The Negroes in our boat were frightened. They had the oars. Duncan tried to talk nicely to them, but it was no good. Nobody could take charge.

I went on bailing water, but it began to rain. Once a Negro kicked me accidentally as I bailed there in the bottom of the boat and I cursed him. "Man," he admonished me—and his voice was that of one who felt he was soon to see his Creator—"man, don't swear here."

At last we saw a light, and a-bending on the oars the Negroes made for it. It was miles away and it drifted off. We had about ten flares and we lit these now and then. Again a light but again it

eluded us. Once we saw a light and it stayed there. We rowed up
to it only to find it was a buoy with some sort of light on it.

All night long we chased these lights. These were not
figments of the imagination but [rather] ships searching for us.
However, they always seemed to go away from us. We never came
up to them.

Morning at last and two ships. They were perhaps ten miles
away. We rowed with all our vigor but it was three hours before
we reached one of them

Once on board the *Berlin* we were treated with the greatest
kindness. I cannot write too strongly of all they did for us. They
gave us clothing and food and cared for us in every way. Captain
von Thülen was most kind to all of us.[60]

Another rescued by the *Berlin* from lifeboat 13 was seaman Joseph
Alexis. Most of the members of the crew were reluctant to discuss
the disaster, but Alexis told how he had helped another seaman pick
up twenty others who were in the water.

I was on the boat deck as the *Vestris* was going down. Boat 13
couldn't be launched and there seemed no place to go. By that
time the sea was covering the deck and I felt the boat settling
under me. I dived into the sea and, looking back after I came up,
I saw the *Vestris* going down.

The sea picked up the boat as the *Vestris* shot downward and
the lifeboat stayed on the surface. That boat was set loose for us
by the Lord Jehovah, I am sure. Several of us swam to it and
climbed in and began picking up others who were swimming
around. When we were picked up in boat 13 on the thirteenth day
of the month, I was sure that must be a lucky number.[61]

Also picked up from lifeboat 13 was Ovelton L. Maxey, twenty-six, of
Richmond, Virginia, who was a steel construction worker. He was
on the same job as the aforementioned Thomas Mack, and was

traveling with him. Maxey and Mack declared, nearly in unison, "The women and children in the boats were murdered as plainly as if they had been hit over the head with a hatchet; the ship's officers waited in sending an SOS just to save a few dollars."[62]

Maxey and others said it required two hours to lower four lifeboats, and Maxey added:

> If the first SOS had been sent out the night before I believe there would have been no occasion for loss of life.
>
> We don't know what caused the disaster, but the vessel was loaded with automobiles and I do not see how the cargo could have shifted. Water entered the coal bunkers from a leak in the hull.
>
> We had a list from 10:00 P.M. Sunday when the gale struck until 10:00 A.M. Monday when the hurricane deck was under water.
>
> We were never told to get ready to take to the boats. The seamen failed to hurry up the lifeboats and the whole thing moved without any precision.
>
> In the lifeboats, we were always in water up to our knees.
>
> We were not frightened; we were prepared to reach a ship, not to die. We knew there had been an SOS sent out. When the *Berlin* picked us up there was a heavy roll but the seas were not very high.
>
> For food we had crackers and water. The crew was fine. If the SOS had been sent out five hours earlier everyone would have been saved.[63]

In an ironic twist of fate, Maxey's brief moment in the limelight as a survivor led to his downfall. On November 17, 1928, he was arrested in his room at the Hotel Woodstock and taken to the prison ward at Bellevue Hospital by a New York City detective on a warrant charging him with embezzlement of $6,750 from a former employer in Cumberland, Maryland. Maryland authorities said they found Maxey through newspaper stories of the *Vestris* sinking and also from photographs of survivors.[64]

Of the first three lifeboats that put off from the stricken liner containing women and children, two sank and one capsized. The *Rockford* (Illinois) *Daily Republic* reported:

§ The terror in the water after the three lifeboats with their precious burdens of women and children had been lost was heartbreaking to the those still on board the sinking ship, but they had horrors of their own to face, for the *Vestris* now began to roll and pitch sickeningly.

Before the rest of the lifeboats could be cleared away the *Vestris* was struck amidships by an enormous wave that rolled it over, capsized. The ship, helpless and hopeless, floated like the carcass of some gigantic dead whale while the passengers and seamen, scrambling for their lives, tried to climb up the slippery base of the keel. Afraid that the ship would sink at any second, they dove from the keel into the ocean, attempting to swim to the nearest boats or to keep themselves afloat with debris until rescuers could arrive on the scene.[65]

Harry Forsyth, the third engineer on the Vestris, was asked if he left the ship or if the ship had left him:

I jumped off her. I had an oar with me from the ship. I found it on the side of the ship as I was walking down the port side of the *Vestris*. I was in the water about an hour and a half before I was picked up by the No. 13 lifeboat, and ultimately by the steamer *Berlin*. There were only four male passengers in No. 13.[†] There were thirty-two or thirty-three persons altogether.[66]

Rescued from the sea: Carl "Iron Man" Schmidt

§ From the *Rockford* (IL) *Daily Republic* of November 15, 1928, came this interesting story of Carl Schmidt, whom they dubbed "the iron

---

[†] Forsyth meant that only four of those in No. 13, all men, were passengers on the *Vestris*; the other twenty-eight or twenty-nine were crew members.

man." Mr. Schmidt was picked up by the German liner *Berlin*. From his account he must have jumped from the *Vestris* just moments before she sank, so call it 2:30 P.M. Monday. He was picked up by the *Berlin* about ten o'clock Tuesday morning.

"I haven't even a cold." Thumping a burly chest with this opening, Carl Schmidt, the "iron man" of the *Vestris* disaster, picked up by the SS *Berlin* after 19½ hours in the water, told the story of his desperate fight for life in the Atlantic.

Some of these hours are mercifully blanks in the memory of Schmidt, for more than once he lost consciousness, succumbing to the grip of the intense cold. The others are chapters of anguish and fortitude such as not one man in a thousand would have lived to tell about. We shall deal with the issue of the "intense cold" later; suffice it to say here that Mr. Schmidt was not reticent about the use of hyperbole. Here is the rest of his story, in his own words:

I waited while they put the women and the children into the boats, and then there was no place for me. When I tried to get into one boat, number 8, they pushed me out. They said it was no good and pointed to a hole in the side. Before I could reach another, all the boats were gone.

Every minute I could feel the *Vestris* sinking lower and lower beneath my feet. There wasn't anything else to do, so I jumped, first making sure my life preserver was well secured.

I swam away from the sinking steamer as fast as I could pull through the water. There were people all around me, forty or fifty of them, all swimming after the boats. They wasted breath screaming for help. I saved mine.

After I had been in the water a few minutes, I reached one lifeboat and made a grab for the gunwale, but I missed. Before I could try again, a heavy sea swept me away. That wave carried me back to the *Vestris* just as she went down. She went down right alongside me.

Here Schmidt stopped to show, with the sweeping gesture of a hand near his side, how close he had been to the sinking *Vestris*.

> She turned over, just like a big potato in a tub of water. Then she disappeared and I had to fight to keep from being drawn into the whirling vortex as she went down.

> Captain Carey was the last man I saw aboard her before she took that final dive. He walked straight over the side as his ship gave her last shudder. He didn't have on a life belt, and that was the last I saw of him.

> Once I got clear of the whirl where the ship sank, I found myself surrounded by passengers, swimming frantically about. Most of them were women, some of whom were trying to keep their children afloat. Lots of them had no life belts.

> Then I went batty.

Mr. Schmidt had no way of knowing how long he was out of his head, but he thinks that his cold-benumbed mind began to work again just before he spotted the *Berlin*:

> I saw the *Berlin* foaming along, crisscrossing back and forth through the wreckage. I almost went crazy again. But it was a combination of joy and fearful anxiety lest I should not be able to attract their attention this time.

> For some minutes I waved my arms as high as I could stretch above the water, and as every minute passed, I felt my heart grow colder than the water all about me. Then at last there came a hail from the *Berlin*'s bridge. I felt warm again. Twice they came up close to me and I paddled toward the side. They had come up on the windward side of me, so that I had some shelter in their lee. But even so, the seas were still high and both times I was washed beyond reach of the two men clinging perilously to the end of a rope ladder.

> The third time, straining out so far that I was afraid they would fall into the sea themselves, they caught me and dragged

me in. I let go then, and I don't remember what happened until I woke up in the ship's hospital.

But a little water couldn't keep me in the hospital. I wouldn't stay. Say, look at me! I'm a big strong man. Forty-five years old and I haven't even a cold. I can stand anything.

When he gave this interview on the *Berlin*, shortly after she docked, Schmidt seemed none the worse for wear, save for some black and blue areas on the back of his neck. He was quick to admit, however, that his back and shoulders were roughly abraded by the rubbing of his life preserver. But he quickly added: "These are welcome wounds."[67]

The *New York Times* added this to Schmidt's story:

Schmidt, a third-class passenger on the *Vestris*, jumped over the side of the sinking ship with a cork life belt wrapped about him and holding two small pieces of wood in his hands. All through the night and until ten o'clock the next morning, Schmidt dog-paddled persistently to the west. He wore a soggy gray cap, under which there were six $500 bills comprising his entire worldly possessions.

Schmidt was spotted just after nine o'clock, but the *Berlin* running under full steam rushed by him before it could be stopped. The big ship took half an hour to turn around and come back to him. He fell senseless when he was pulled on deck.[68]

As for the "cold water," contrast Schmidt's description with that of Paul Dana, related earlier: "The water was warm—warmer than you find it in an indoor swimming tank."

Other survivors told similar stories. Almost all of them, however, said that the air and wind were quite cold. Some of them said they tried to stay in the water as much as practical to avoid the chill of the windy North Atlantic weather.

There were also reports that the *Berlin* had picked up a man clinging to some wreckage; his name was not given. Captain von Thülen of the *Berlin* told reporters this:

> Toward dawn we saw another boat, swamped and drifting with the wind. Moving to the windward, we lowered a Jacob's ladder and Chief Officer William Dähne and Second Officer E. Methy went down. Near the boat was Carl Schmidt. Another man was in the lifeboat, and he was snoring, so exhausted that he was not awakened when carried aboard. The boat was one-third full of water and there was a driving rain.[69]

The sleeping man was identified as Robert Chase, a seaman from the crew of the *Vestris*. The captain's report implies that he was alone in this lifeboat. Which lifeboat this was remains a mystery.

As for the mystery man who was "clinging to some wreckage," all we have is this from a story in the *Kingston* [NY] *Daily Freeman*:

> Fragmentary stories of the horror on the descent from the sloping decks of the *Vestris* into a rough sea and the afternoon and night adrift in the small lifeboats have come from survivors. One man, who was rescued by the *Berlin* from a floating spar to which he had clung through the night, told of seeing a woman and child adrift in a lifeboat. Although the *Berlin* put about immediately and retraced the course the man was believed to have floated, no trace of the woman or child could be found.[70]

≈ ≈

Rescues by the Battleship *Wyoming*

The battleship USS *Wyoming* was a latecomer to the rescue effort, steaming onto the scene around daybreak on Tuesday. She was under the command of Captain Luther M. Overstreet, and her navigating officer was Commander G. W. Simpson of New York.

Four survivors were picked up who had been clinging to the overturned lifeboat No. 8 all night long, and four more were rescued from a raft or clinging to wreckage.

The *Wyoming* rescued a total of eight persons from the angry sea that day: Teruko Inouye, wife of the Japanese military attaché to Argentina; Dolores Barreiro Doril of Brooklyn, New York; Marion C. Batten of Altoona, Pennsylvania; Marie Ulrich of Brooklyn, New York; Elvira Fernandez Rua of New Bedford, Massachusetts; and three firemen from the *Vestris*: John Morris, Gerald Burton, and Joseph Boxill.

Soon after daylight Tuesday, the *Wyoming* sighted the *Myriam* and the *American Shipper*. Commander Simpson told reporters of the *Wyoming*'s part in the rescue:

> [The *Myriam* and *American Shipper*] reported at this time that all but two lifeboats—believed to have been swamped—had been accounted for. Realizing that the raft would not drift as fast as a lifeboat in the wind, the *Wyoming* shaped her course into the wind and at 8:30 A.M. we sighted wreckage. The *Wyoming* so maneuvered as to bring this wreckage under her starboard bow. There was a man on the wreckage. A heaving line thrown overboard to the lone occupant of that piece of wreckage hit him on the arm. He made no move. We lowered a man overside with a line, and hauled the man aboard. He was just about unconscious—[Joseph] Boxill, stoker.
>
> At the same time our lookouts reported men and women swimming in the water from dead ahead to the full beam of the *Wyoming*, as well as four persons on an improvised raft to the starboard.

We lowered boats and picked up these people. The four individuals rescued from the water were Mrs. Ulrich, Mrs. Batten, John Morris and Gerald Burton. Their boat had been swamped. After these four were brought aboard, our motor lifeboat approached the improvised raft on which were four persons. Three were alive: Mrs. Inouye, Mrs. Rua, and Mrs. Doril. The man was dead. Mrs. Inouye and Mrs. Batten told us afterward that their husbands had died in their arms. Major Inouye, we were informed, lived until two hours before his wife was rescued; she said she continued to cling to him as long as she could after he had died because she did not care to leave him.

While these rescues were being made our lookouts kept reporting other bodies floating by. They were examined in all cases and not taken aboard unless signs of life were visible. This was considered necessary because we were trying first to save the living in a limited time and a wide area had to be covered. We saw about twelve dead bodies, both men and women. It was a ghastly sight.[71]

Joseph Boxill, first rescue of the *Wyoming* and a fireman on the *Vestris*, told his story in the federal inquiry conducted by District Attorney Charles H. Tuttle, giving a narrative of his whole experience, including twenty hours in the ocean without a life belt:

§ It was then after eleven, he said. Despite further increase in the water, the port boiler was still being fired. Shortly after, all but he and a companion [fireman] named James Oliviere left the stokehold. He tried to get Fifth Engineer Jones to muster the others, but Jones was able to do nothing, and simply told him to carry on. Returning to the stokehold, Boxill found that Oliviere also had left, and he gave up and went on deck.

Above he found all the crew had donned life belts, so he put on his shore clothes and a life belt. He went to his boat station at lifeboat 4. Boxill was in No. 4 when it was being lowered. He and another member of the crew had knives ready to cut the tackle, but the boat was still some distance out of the water, and the order to cut the ropes never came. Instead, the *Vestris* rolled over on her side, and the boat lay tilted on the hull of the ship, spilling some of the passengers out. He himself jumped out, threw off his life belt, and swam off some distance.

Continuing his story, Boxill testified directly:

I found a piece of wreckage, and then I found another, larger piece. Then I saw a boat that was three-quarters full, and swam toward it. But just before I got there it turned over. I held onto the side, and a man tried to turn it back on its keel so that the passengers could stand inside it. [Ed. Note: Probably No. 8.]

There were some dead children and women around there, sir, and that got me kind of dizzy, so I pulled my clothes off there; I thought it would be a good place to leave my clothes, and swam off in my combination suit.[†] I found a good large piece of wreckage, I think it was a plank from the salt water bath. Then I swam until dusk, when I saw the engineers' messman on a large piece about six feet by four feet. I stayed on top until it was dark, and then there was a larger piece, a raft. There was another man, a passenger, on this, and I said, "Slim, let's swim over to that thing."

But Slim couldn't swim, so I had to swim and tow the piece we were on and try to get it so that Slim could get on the raft. It was pretty hard work, but I paddled down to the wreckage and the three of us stayed on that.

In the night, Slim cried out from a cramp. I don't know if it was midnight, but it was terribly dark, and I could judge from that. Slim couldn't hold on any longer, and he slipped off. After that the passenger floated away, too.

[†] An undergarment similar to long johns.

Then I saw a red flare way up in the air, just as I was about to
give up. It seemed a million miles from there. Then I knew
nothing more until half past twelve in the afternoon, when I
woke up on board the *Wyoming*.

Boxill was asked what became of the women and children in boat 4,
and he replied that "some got thrown out of the boat." No officer
ordered the occupants out of the boat while it was hanging helpless
at the side of the ship, and the last order that Captain Carey gave
was, "Lower away."[72]

The *Wyoming* next rescued the survivors of the doomed No. 8
lifeboat. Only four still clung to the overturned boat. When they saw
the Wyoming draw near to them, they let go of the overturned No. 8
and swam to the *Wyoming*.

Gerald Burton, "leading fireman of number one watch," who
gave his address as Bay Street, in Barbados, West Indies, said he did
not see the hole in lifeboat 8. The boat filled and capsized six times
before the four were picked up at ten o'clock Monday morning.

Burton surmised that lifeboat 8 was "bilged"[†] by the *Vestris's*
starboard [bilge keel] as she was being lowered into the sea from the
davits. [Bilge keels], normally under water, are shelter-like plates built
out from the turn of the hull and extending all along it to steady vessels in a beam sea. Burton said he had not seen anyone nail a tin plate
over the hole in No. 8.

This lifeboat was in command of the *Vestris's* barman,
Burton related. It was the boat that Second Officer Leslie Watson
should have had charge of.[73]

Burton's opinion about the hole in lifeboat 8 was backed up by
testimony from Edward Walcott at the Custom House hearing

[†] Bilged: condition of a vessel in which water is freely admitted through
holes and breaches made in the planks of the bottom, caused by damage to
the bottom of the boat.

under Inspector General Hoover: Edward Walcott identified the lifeboat that was holed as No. 6, not No. 8. He said that it occurred during the lowering of the boat and that it appeared to have been stove in.[74]

Gerald Burton was hailed as a hero by some of the survivors.

"If it hadn't been for Gerald we wouldn't be here," was the comment Marion Batten made just before she lapsed into sleep at the naval hospital.[75]

Marion Batten's husband was Indy race car driver Norman K. Batten. They had been traveling with Earl and Anne DeVore; Earl F. DeVore was also an Indy car driver. The two men were to participate in a special racing event in Argentina.

The *New York Times* interviewed Marion Batten in her room at the Hotel Belmont in New York City.

§ Marion Batten, who had been in lifeboat 8, spoke in a low voice as she related her night-long struggle in the ocean to keep her husband's head above water, and of how, with help at hand, a wave had wrenched him away to drown. She did not let emotions get the better of her until she reached the end of her story.

Speaking slowly and unemotionally, she said:

At 6:30 Monday morning my husband awoke me. There was a bad list. He dressed hurriedly and went up on deck to find out what was wrong. I dressed as quickly and went to the smoking room, where I had a buttered roll. That was the only food we had on the *Vestris* that day.

We got life preservers and went out on deck. After much delay we got into a lifeboat. We got into number 8. There was a hole in it, and it began to leak right away. No sooner had we left the *Vestris* than the ship went down. I was facing the other way, and when I looked around it had gone. There were a hundred people floating around.

We tried to row, but the sea became very heavy and it practically swamped us. My husband was very exhausted from trying to keep order and bail.

Mrs. Batten was asked what she meant by "keeping order." She explained that he tried to "make more people work at the bailing." Boat No. 1, with "four or five of the crew in it," had taken her friend Anne DeVore, and she herself had been about to move to the other boat.

There were sixty in our boat, and they had only a few. I had my foot in the other boat, but a Negro pushed me back with an oar to my chest. He said, "We don't want no more women in here," and [they] moved away.

After an hour our boat filled up and it turned over. We righted it and clung on. The day was very long. Then the night came. There was the horrible blackness. There was thunder and lightning and heavy rain. At last we saw two searchlights, and that gave us a great deal of hope. But they didn't see us.

Every time the boat turned over we would hold on. One by one the others seemed to be disappearing. It was a long time before dawn. When it got light there were ten of us left—ten out of sixty. Mrs. Ulrich, a German woman, was the only other woman. We figured that the searching parties would reach us soon.

My husband was growing weaker and weaker. He had exhausted himself bailing. He was delirious at times. I tried to hold him up.

Mrs. Batten did not give details of her struggle to save her husband's life, but others told of her holding onto him every time a wave loosened his grip, and of resting his chin on a life preserver in her frantic efforts to keep his head above water.

We saw two vessels only about a mile away. We raised a piece of cloth on an oar and waved frantically but it was no use. A huge wave washed my husband away from me. The last I saw of him he was floating away with his face down. That was the worst moment of the whole [paused] . . . the whole tragedy.

I felt I could not hold out myself. I sent an SOS to God. "If help is coming, send it quickly, God," I said. Then the *Wyoming* appeared and gave the remaining six of us new hope. We raised our flag, and then we broke away and swam. I tried to swim somehow. They sent off a lifeboat and picked us up. They gave us marvelous care.

When she was asked to describe the demeanor of those in the lifeboat when they knew they were sinking and that empty boats would give them no help, Mrs. Batten showed the emotions that these memories evoked in her. She said that there "was a lot of cursing of the crew and the captain. The captain got most of the curses. At ten o'clock Monday morning my husband asked the captain if he had sent an SOS. He did not reply directly, but said that he was worried, and a little later we found he had sent an SOS."

During the night, she said, while those clinging to the water-logged lifeboat were dropping away one by one, "there was no conversation—only prayer."

Her thoughts went back to her husband and she said, "He passed away easily. He was delirious. He made a desperate last clutch when the wave took him."

A woman who was listening remarked: "Then you had no chance to say good-bye."

Mrs. Batten's lips opened to say "No," but she quickly covered her face with her hands and bending forward, wept quietly.[76]

§ Three persons survived on the improvised raft and were the last survivors picked up by the *Wyoming*. They were Elvira Rua, Dolores Doril, and Teruko Inouye. They were taken to New York by train.

Elvira Rua, who lost her husband and her two-year-old son, wore wrinkled clothing when she arrived at the train station from Norfolk.

Her left hand, with three fingers broken, was in splints and both legs were bandaged. She cried hysterically as she was helped from the train. She left for her home in New Bedford by automobile.

Dolores Doril appeared dazed. Seeking a familiar face in the crowd at the train station, she espied her cousin Lola Doril, and inquired excitedly after her husband, Juan Doril, who had been rescued by the *American Shipper*. He told a story of the disaster from the viewpoint of one who had spent thirty-five years at sea.

"I am not a licensed engineer," said Mr. Doril, "but after spending twenty-five years in engine rooms I know that the *Vestris* sank because the coal port was open. That is not an opinion—I know." He said he saw water in the coal bunkers at 2:00 A.M. Sunday.

Major Inouye, Japanese military attaché to Argentina, and his wife, Teruko, were part of a group of passengers from lifeboat 8. Through a translator, Teruko explained that they had been thrown into the water as their lifeboat sank, and that her husband had died as she clung to a makeshift raft (consisting of debris from the ship and life preservers) with one arm, and to her husband with her left arm. After Mrs. Inouye lost consciousness on Tuesday morning, Elvira Rua and Dolores Doril helped keep her and her husband's body afloat.[77]

After the disaster, Teruko Inouye gradually lost use of her left arm with which she had supported her husband. In 1938 she held a small *Vestris* reunion in New York City, to which she welcomed several crewmen whom she felt had behaved most heroically.[78]

Of thirteen children who were aboard the *Vestris*, not a single one survived. Out of thirty-six women passengers, only eight were rescued. Yet fifty-two of the seventy-nine male passengers on board lived. A sorry record by any standards.

~ ❀ ~

The Battleship USS *Wyoming*, ca 1912-13

This ship arrived at the place where survivors had been picked up about daybreak on Tuesday, November 13, 1928. She picked up eight more persons from the shipwreck. The *Wyoming* was commanded by Captain Luther M. Overstreet, and her navigating officer was Commander G. W. Simpson. The Battleship USS *Wyoming* was commissioned September 25, 1912, and was the lead ship of her class.

Captain William J. Carey (right) and Chief Engineer James A. Adams

Captain Carey was harshly judged for the *Vestris* disaster, but he was a man caught between two deeply conflicting obligations. On the one hand he had his loyalty to the shipping line that employed him, and on the other hand he had his responsibility for the ship's crew and passengers. He was unable to resolve these conflicts in time, and so he went down with his ship in the time-honored tradition of the seafaring man. Adams survived the wreck.

Artist unknown / *Baltimore News*

Artist's conception of the sinking Vestris

This drawing, published in the *Baltimore News* for November 15, 1928, shows the *Vestris* capsized and sinking rapidly as last-minute persons make a desperate scramble for their lives. The drawing is signed by the artist, but the name cannot be made out from the picture, which exists only on a few Internet websites. Although made from whole cloth, it depicts the horror of the scene quite adequately.

Photographer unknown / New York *Daily News*

Teruko Inouye is taken ashore at Hampton Roads, Virginia.

Madame Inouye, now the widow of the Japanese military attaché to Buenos Aires, Yoshio Inouye, was rescued from a makeshift raft by the battleship USS *Wyoming*. Here she is being transferred from the USS *Wyoming* to an auxiliary boat at Hampton Roads, Virginia, on November 15, 1928. Mrs. Inouye held her husband up for many hours to prevent him from drowning but he slipped under the water just two hours before she was rescued.

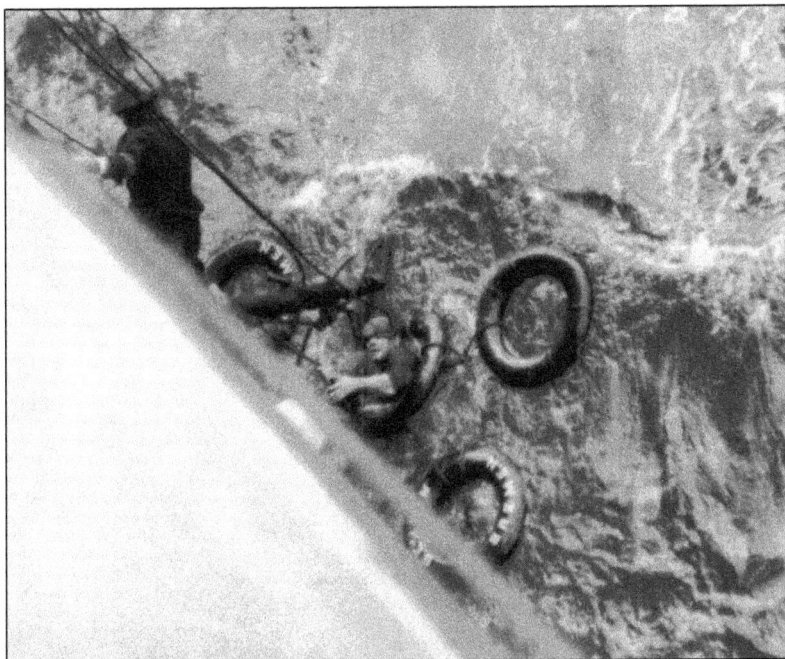

Rescue of Carl Schmidt by the SS *Berlin*

This dramatic photo shows Carl "Iron Man" Schmidt being picked up by the *Berlin* from the roiling sea about ten o'clock in the morning of Tuesday, November 13, 1928. As Schmidt described it:

> Twice [the *Berlin*] came up close to me and I paddled toward the side. They had come up on the windward side of me, so that I had some shelter in their lee. But even so, the seas were still high and both times I was washed beyond reach of the two men [who were] clinging perilously to the end of a rope ladder.
>
> The third time, straining out so far I was afraid they would fall into the sea themselves, they caught me and dragged me in.

The *Voltaire* leaving its pier at Hoboken, New Jersey

Bound for Barbados, the SS *Voltaire* leaves port in Hoboken, New Jersey, carrying some of the sixty passengers who survived the sinking of her sister ship, the SS *Vestris*, just twelve days earlier. Photograph was taken on November 24, 1928, by *Daily News* photographer Leonard Detrick.

Former opulence of the steamship *Vestris*

This picture, taken from a Lamport & Holt brochure, shows the first-class music drawing room as it originally appeared in 1912.

Four officials involved in the *Vestris* hearings

Left to right: Captain E.P. Jessup, representing the American government as a nautical expert; US Department of Justice Commissioner Francis A. O'Neill, before whom the hearings were held; Captain Henry McConkey, marine Superintendent of the Cunard and allied lines in New York, who acted as a nautical expert for the British government; and District Attorney for Southern New York Charles H. Tuttle, who questioned the witnesses.

Note the Phi Beta Kappa keys on both McConkey and Tuttle.

Chapter Seven
Aftermath

THE RIPPLES FROM THE SINKING *Vestris* had hardly faded away before the accusations and controversy exploded. People everywhere were astonished that such a thing could have happened.

The question on everyone's mind was: Why was the SOS sent so late? The answer went to the bottom of the Atlantic Ocean with Captain Carey. Now he was vilified as incompetent, a fool, or—even worse—a murderer. Yet at age fifty-nine he had a stellar record as a ship's master dating back to 1914. He served for forty years on Lamport & Holt vessels. During the Great War he served on troop transporters and supply vessels, surviving a torpedoing of the *Titian*.

Despite all of this, many survivors of the disaster believed that Captain Carey had orders from the company he worked for to avoid the salvage costs that would be incurred by sending an SOS. [$]An unnamed psychiatrist was reported by *Time* magazine as having inferred that the crisis had rendered Carey bewildered, overwhelmed, and mentally paralyzed.

The truth may be that Carey was simply misled by bad information fed to him by his officers and believed that an SOS was not necessary—until it was too late.[1]

IT WAS INEVITABLE that court hearings and investigations would be made into such a disaster. The first inquiry was initiated by US Department of Justice Commissioner Francis A. O'Neill, who held hearings under the direction of the District Attorney for Southern New York, Charles H. Tuttle. We refer to this investigation as the "Tuttle hearings" in this book. The hearings began on November 15, 1928, when Attorney Tuttle took the testimony of six survivors.

At the close of the first day's hearing, Tuttle remarked:

§ "From the testimony taken in today's hearing before Commissioner O'Neill, particularly that of the excellent witness Fred W. Puppe, electrical engineer, it is fully apparent that we shall ultimately obtain evidence to fix blame for the wreck and the great loss of life."

On the second day of the hearings, Chief Engineer James Avard Adams testified, "As late as 11:00 A.M. Monday I reported to Captain Carey that the *Vestris*'s pumps could keep her afloat another six hours, when Captain Carey expected US destroyers to arrive."[2]

§ In addition to the Tuttle hearings at least seven other organizations planned their own investigations into the loss of the *Vestris*. They included the following:

(1) The US Department of Commerce under Dickerson Naylor Hoover, Supervising Inspector General of the Steamboat Inspection Service; to determine if the *Vestris* was inspected properly in port.

(2) The Commerce Committee of the US Senate, Senator Jones of Washington, chairman.

(3) The Board of New York Insurance Underwriters.

(4) The Central Trades and Labor Council of Manhattan, to determine if the crew of the *Vestris* was underpaid or incompetent.

(5) The British Board of Trade (which hearing did not occur until April of 1929).

Captain Bambra testifying at *Vestris* hearing

Captain William Andrew Bambra (right) testifying during the federal inquiry into the *Vestris* disaster late in November 1928. Captain Bambra was a former master of the *Vestris*. The "ghost" in the foreground is the head of somebody who moved during the fairly long exposure needed to take this photo without the aid of a flashbulb, which likely was not allowed.

(6) The British Parliament.

(7) Sanderson & Son, agents for Lamport & Holt.[3]

It sounds as though everyone wanted to get in on the action, in contrast to the *Titanic*, which had only two major inquiries.

At the Tuttle hearings, former *Vestris* master, Captain William Andrew Bambra was questioned by nautical expert Captain E. P. Jessup, who asked Bambra how he would act in a situation such as that in which Captain Carey found himself. He replied that Carey, whom he had known for a long time, was an excellent seaman with a clean record and an excellent reputation. Bambra did not actually address the question he was asked. When asked how much water would be required to give the ship a list of thirty-two degrees, he replied, "hundreds of tons, possibly a thousand tons." Capt. Bambra gave amounts of cargo and stores carried by the *Vestris* that were at odds with those given by other witnesses.[4]

The "Ship's Nostalgia" website had this additional account:

§ The American press, with considerable justification, were highly critical of the incompetence of the master, officers and crew of the ship and the management of Lamport & Holt. This led to a dramatic drop in loadings for the company's other liners and the South American service was discontinued at the end of 1929. The other ships employed on the service were brought back to [the] UK and laid-up.

Wikimedia Commons

The *Celtic* aground on Calf Rocks, County Cork, Ireland

Loaded with surviving crew members of the *Vestris*, she was a total loss, but all persons on board were saved.

The British *Vestris* survivors were brought home on the White Star liner *Celtic*. As she approached Cobh on 10 December [1928] the ship stopped to pick up a pilot in gale force conditions and drifted onto Roches Point at the entrance to the harbor. Full astern was ordered and *Celtic* came off but went aground again on Calf Rocks and became a total loss, thankfully without loss of life.[5]

The survivors were all British crew members of the *Vestris*.

A matter of dead reckoning

The ships coming to the aid of the shipwrecked occupants of the *Vestris* took a long time to arrive. Survivors told of seeing searchlights on the horizon for many hours before help actually arrived. Why was this? It turns out that the pivotal error was made by not taking into account the drift of the debris field of the sunken *Vestris*.

The debris field—the lifeboats and floating wreckage of the *Vestris*—was not in a fixed position but was rather a "moving target." It drifted with the Gulf Stream at an eastward rate of about five knots. When the first rescue ships to reach the SOS position arrived about 7:30 P.M. Monday night, the debris field had been drifting for some five hours, putting it roughly twenty-five miles east of its original position. They were looking in the wrong place.

§ In testimony given to the federal board of inquiry, Captain Luther M. Overstreet, the commander of the US Navy battleship *Wyoming*, which picked up six survivors, related the reasons for the unfortunate miscalculation. He testified that nine ships were at the position given in the SOS "several hours" after the *Vestris* sank.

Searchlights were used in the search, but Overstreet said that only vessels approaching from the east—the *Berlin*, the French tanker *Myriam*, and the *American Shipper*—reached the area where the lifeboats were found. Three other ships were near the SOS position, but the lifeboats were actually found 37 nautical miles to the east of that position.[6]

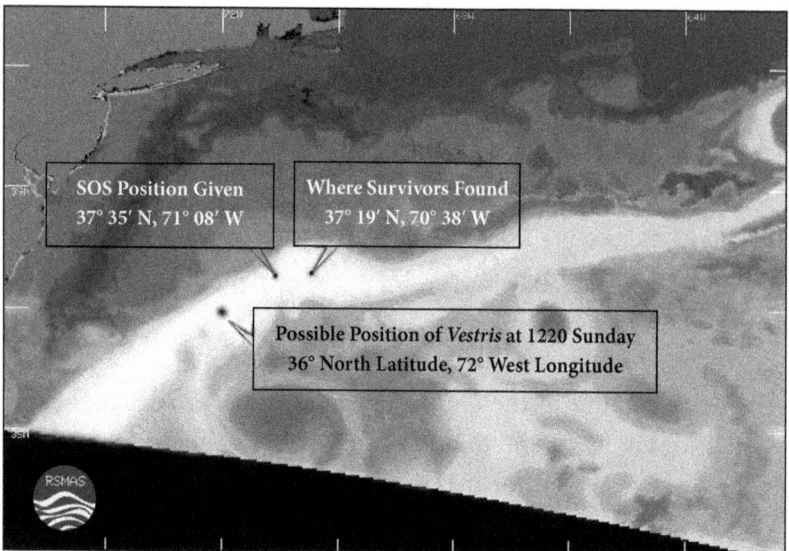

Gulf Stream map showing some probable locations for the *Vestris*

The *Vestris* left Hoboken at 3:45 P.M. on Saturday, November 10, 1928. At her rated 15 knots (nautical miles per hour), she should have made about 300 nautical miles (n.m.) when she was hove to at 12:20 P.M. on Sunday the 11th, and she may have been carried about 100 nautical miles north-northeast by the Gulf Stream before sending the SOS. The lowest of the three black spots—the fuzzy one—is the result of these two vectors. At 2:36 P.M. the next day, Monday the 12th, *Vestris* sank, after having sent out an SOS at 9:56 that morning giving her position as 37° 35' North, 71° 08' West. This is believed to have been "substantially correct," but subject to a possible error of up to ten nautical miles in any direction. By the time rescue vessels found survivors at 3:25 A.M. the next day, the debris field probably had drifted another 30 n.m. to the reported location of 37° 19' N, 70° 38' W (subject to a possible error of ± 5 n.m.). When she sank, the *Vestris* was about 187 nautical miles from the nearest land at Atlantic City, New Jersey.

This infrared scan of the West Atlantic was made by a Moderate Resolution Imaging Spectroradiometer (MODIS) satellite.[7] The warm waters of the Gulf Stream are shown in white, and the cold waters of the Labrador Current are shown in grays and black. If the Gulf Stream was located as it was in this scan, the *Vestris* would have been in the middle of the current.

Who knows how many passengers and crew members of the *Vestris* died because of this simple yet avoidable error in tracking a moving debris field?

The map on the previous page shows the location of the *Vestris* and her distance from land at the time of sinking.

We have already given some of the testimony of early witnesses in the Tuttle hearings. More details emerged as Attorney Charles H. Tuttle became more insistent in his questioning of witnesses.

§He first obtained from James MacDonald, assistant radio operator, an admission that Michael O'Loughlin did not go down with the ship, calling for help to the last, as had been stated earlier by other witnesses. MacDonald said that he left the radio cabin with O'Loughlin. He said they left the radio room maybe a half hour or perhaps even forty-five minutes before the *Vestris* sank, but in all of his testimony, MacDonald was vague about times.

Tuttle then questioned Charles Verchere, second assistant radio operator, who appeared nervous and chewed on his fingernails during his testimony. He told Attorney Tuttle that he did not know the exact time the SOS went out,[†] but he thought a CQ[‡] message (an appeal for ships to stand by) was put on the air about nine o'clock Monday morning.

Dealing with his last minutes on the *Vestris*, Verchere told Tuttle that he went to the wireless room and warned O'Loughlin and MacDonald to get out as quickly as possible, as some of the boats had left, and if they didn't leave at once there would not be much chance of doing so at all.

[†]At the British Board of Trade inquiry, Verchere testified that the first SOS was sent out at 9:56 A.M. EST Monday morning by O'Loughlin on the instructions of the first officer, Mr. Bolger.[8]

[‡] The CQ distress code was originally CQD, meaning "Come quick, danger."

The last he saw of O'Loughlin, Verchere said, was that he was walking along the deck; Verchere didn't know where he was going.

On Thursday Attorney Tuttle was sympathetic to his witnesses, all of whom were passengers on the *Vestris*. He let them tell their stories in their own way, at times asking a quiet question. Friday he was bellicose; he put questions in a sharp, harsh voice and paced alongside the witness stand. He was particularly exasperated with Verchere, the radio man, and gave him the third degree.

Typical of the exchanges between Attorney Tuttle and Charles Verchere was this one:

"Let me understand it completely," queried Tuttle. "Were you present when the SOS call was sent out?"

"Yes," Verchere answered.

"Was there a clock or a chronometer in the room?"

"Yes," replied Verchere, "there was a clock with Greenwich Mean Time on it."

"With Greenwich time on it?" asked Tuttle. "Yes," was the reply.

"Did you assist in sending that SOS?"

"Well," asserted Verchere, "it had nothing to do with me."

"You were present when it was sent out?" asked Tuttle.

"Yes," affirmed Verchere.

"And you did not notice the time at which that thing was sent out?" Tuttle asked, as if he found that hard to believe.

"There was no clock in the wireless room," replied Verchere.

"You just said a moment ago. . ." began Tuttle, but Verchere cut him off with the statement, "That had Greenwich time on it."

"Wasn't it going?" asked a wide-eyed Tuttle.

"It had Greenwich time on it," repeated Verchere.

"What of it?" retorted Tuttle.

"I did not take any particular notice of the time, no," asserted Verchere.

Tuttle nearly exploded. "Do you mean that the sending out of an SOS upon which the lives of everybody on the boat depended, and you being one of the radio operators, you did not notice at what time that was sent out? You want us to believe that story?"

"About ten o'clock I should say," replied Verchere.

"Yes, but you said a little while ago, you would say very roughly?"

"It was roughly," agreed Verchere.

"I want to try to pin you down in something," Tuttle said.

"I could not give you any definite time for anything."

"Is that," asked Tuttle sarcastically, "a formula you have prepared before you went on the witness stand, that you would not be definite about anything?"

"No. You can't expect me to have a mind very clear after fourteen hours in a boat, can you?"

Tuttle then excused Verchere and called the other surviving radio operator, James MacDonald, to the stand.

"The list was beginning when I went on duty Saturday midnight," he testified. "When I got up at eight o'clock in the morning of the 12th it was pretty considerable."

MacDonald told how he had escaped from the sinking ship:

I was working the key, and I said, "Well come on you two, there's nothing more we can do." So O'Loughlin went to the key and sent out the message, "Abandoning ship." O'Loughlin and Verchere went out on the high side of the ship and I went on the low side. I slid down to my boat. Johnson was in charge.

Now our boat was very near the water, because the ship was rolling. [Johnson] said, "Now when I say pull you pull, and at the end of the boats there are two handles which disengage the hooks, and the boat just goes down like that in the water." He was watching his chance. I said, "You hold onto the side because maybe she is bumping and she will shake you out."

Mr. Johnson said "Pull," and the fellows half pulled and he started cutting the rope. He was too slow and the boat went down like that, and the water rushed up.

Because the water was above the gunwale of the boat, the oars floated out of the oarlocks and it was only with great difficulty that we worked the submerged boat away from the side of the ship before she went down.

We looked down the ship's side and there were two boats hanging there. There was a boat with only four in it, and we were going to it, but when we saw the two boats on the side of the ship we waved to the other boat to go and help them. Then I got very sick, and I was attending to myself and did not see the ship go down.

After a while another boat came by, and the chief officer and half the crew jumped off and swam for that boat. I thought I would go out with the next boat. But the next boat was No. 13 and I didn't like that number, so I said I would wait a little. I saw another boat coming and I swam to it, and they pulled me inside. Verchere was on that boat.

MacDonald brought the first bit of humor to the hearing when he told how he unknowingly sat on the water can in his lifeboat and kept many persons from drinking water for hours.

"No one ever tried to open the water keg," he said, "I am certain of that because I had been sitting on it all night."

Women who had been leaning against the walls of the band-box room, weeping quietly into their handkerchiefs, dried their faces and smiled as the red-haired Scotsman went on to recount how the men in his boat, unworried by the perils of the open sea, yelled for cigarettes. Even Commissioner O'Neill chuckled while the United States marshals shouted for order in the court.

Arthur J. Costigan of the Radio Marine Corporation brought copies of three radiograms to the hearing, one of which was sent by

Captain Carey to Lamport & Holt and transmitted by the Radio Marine Corporation, which read:

"Hove to since noon yesterday. During night developed thirty-two degree list. Starboard decks under water. Ship lying on beam ends. Impossible to proceed anywhere. Sea moderately rough. Carey."

Attorney Tuttle then introduced a theme that would run through his interrogations for the next few days.

"I want to say at this time," the federal attorney said, "that if any person is found guilty of trying to influence witnesses or remove them from my jurisdiction, he will be punished. We want the facts and we must have witnesses to get the facts."

The fourth witness of the day was Alfred Hanson, an assistant pastry chef on the *Vestris*, who found time to get out his camera and take pictures while carefully moving around on the sloping deck of the vessel from nine o'clock Monday morning almost to the time when the *Vestris* sank. He was on the stand only long enough for Mr. Tuttle to introduce the pictures as exhibitions.[9]

Evidence of witness tampering

The idea that someone was tampering with witnesses was brought out even more by other testimony.

§Attorney Tuttle's severe examination of the assistant radio man, Charles Verchere, reportedly had the young Scotsman squirming in the witness stand like a boy called into the principal's office. Tuttle pushed him to admit that a message had been sent Sunday to Lamport & Holt that the *Vestris* was in trouble. He started by asking:

"Prior to the time when the SOS was sent out was there any message sent out by that vessel except private messages for the passengers and this particular government message? Take your time and think about it and give me a truthful answer."

"There may have been a message sent, but I don't remember. I didn't send that and I don't remember."

"Downstairs didn't you tell me that there was a message sent stating 'We may need aid'?"

"Yes. I am not going to say anything about that message because I don't remember about it."

"Well, now, let's go into this message that you know was sent, but don't remember about. Who told you not to remember about that message?"

"Nobody."

"Can you tell me from where you learned about this message, from what source? Did you learn about it from reading it?"

"What I thought was that there was a message sent just before the SOS very shortly, on that Monday morning."

"Mr. Witness, you have testified five minutes ago that that message may have been sent before you came on duty at eight o'clock Sunday night. You testified to that only five minutes ago."

"It may have been sent any time, I was not on duty and I would not have known about it."

"But you told me all about it down in my office; why have you forgotten now?"

Tuttle then addressed Commissioner O'Neill, saying, "If your Honor please, I think you appreciate now why it was that I made a certain statement."

"I do," replied O'Neill, and addressing the witness he asked, "Young man, you appreciate you are under oath?"

"I do, yes. Yes, sir. I know. I have not said anything that I do not know and that is not right. I cannot tell you about the message."

The "certain statement" Tuttle was referring to occurred at the beginning of the day's hearing. In his remarks to the court at the opening of the session, he brought up the possibility that "persons unknown" were attempting to obstruct the investigation and added that he had information from "confidential sources" that such obstruction was happening.

"Surely," he said, "the Court will deal severely with anyone who undertakes to prevent the revealing of the entire truth in this investigation, either through some association with particular witnesses or by removing witnesses from the jurisdiction of the Court."[10]

It is impossible to read accounts of the federal hearing—of which those given here are merely a sample—without coming to the conclusion that Lamport & Holt officials were complicit in witness tampering.

Inspector filed false report on the *Vestris*

§Another surprise arose when Edward Keane, inspector of hulls in the United States Steamboat Inspection Service of the Department of Commerce, admitted at the Tuttle hearings that if he had made a truthful report of his inspection of the SS *Vestris*, it would not have been issued clearance papers.

At the Tuttle hearings, Keane reiterated testimony he gave first at the inspection service's investigation that although in his official report he said that he had lowered the *Vestris*'s lifeboats, he had not actually done so.

"That being a requirement," asked Attorney Tuttle, "unless it was done clearance papers should not be issued, should they?"

"No," the inspector agreed.

Tuttle produced Keane's official inspection report and read from it a printed question as to whether the lifeboats had been lowered. Next to this question Keane had written "yes."

Keane repeated his earlier testimony that he had not lowered the boats because at the time of the inspection the dock was on one side of the *Vestris* and barges on the other. If he attempted to lower the boats with a full load, in accordance with the rule, he said that "something might give way and a disaster result." He was certain, however, that the lifeboats and their gear were in excellent shape.[11] [Ed. Note: If he didn't try them, how would he know?]

Officer says *Vestris* was not seaworthy

§ Only the day before the latter testimony, on November 27th, Harry Wheeler, superintendent of the Lamport & Holt Line, operators of the *Vestris*, testified that if the *Vestris* put to sea on her fatal voyage with no covers on the hatches, she was not seaworthy.

Testifying at the Tuttle hearings, American nautical expert Captain Jessup asked if such covers were provided on the *Vestris*.

They were not, and Jessup added, "If the sea was coming up, was it not the first duty of a captain to see that covers for these hatches were improvised?" Wheeler answered that it was.

> Reginald Dickson, sixth engineer on the *Vestris*, testified at the British Board of Trade inquiry that on Monday morning the colored firemen left the stokehold.
>
> When he was asked, "All of them?" Dickson answered, "No, most of them."
>
> "The captain told us to get together as many firemen as we could." When asked if they got the firemen below again, Dickson replied, "Yes." Asked if they stayed below, Dickson replied, "No, they walked into the stokehold and went up again."[12]

Dickson's testimony certainly explains why the ship's officers were not on hand to supervise the launching of the lifeboats.

Crew member denies "mutiny"

The Topeka, Kansas, *Plaindealer* reported this:

> The stories of mutiny on the part of the firemen were refuted by coal trimmer Joseph Garner, who flatly denied charges of "mutiny," asserting that the crew obeyed every order and stuck to their posts, though they had to work in water and without food for two days, and that all were calm when the ship sank, which really surprised him.[13]

The bosun found he could not open this door because the bolts stuck; they were rusty and had not been greased

The stoke hole began to flood with water—tons of water—and it kept coming into that stoke room too fast for the pumps

Artwork by International Illustrated News in the *Augusta Chronicle*

Artist's conception of stokers fighting the flood

We were working in water waist-high and were knocked off our feet with each roll of the ship . . . we rigged lifelines, to which firemen were tied in single file, so that we could pass sacks of coal from the bunkers

In the night the pumps choked up and stewards and firemen were bailing with buckets ... we stayed there until the last minute, all of us working as hard as possible—but it was useless

Artwork by International Illustrated News in the *Augusta Chronicle*

Artist's conception of stokers fighting the flood—continued

The situation that confronted the stokers is illustrated by the crude drawings on the previous two pages, which were found in the *Augusta Chronicle* and originally captioned as "How *Vestris*'s Stokers Battled Rising Waters." They are an artist's depictions, obviously, but they show rather well the extreme circumstances under which the stokers were working in those final hours.[14]

The testimony of First Officer Frank Johnson

By November 17, 1928, evidence of obfuscation on the part of Lamport & Holt began to appear. [§]After dodging subpoena servers for forty-eight hours, Chief Officer Johnson, who was second-in-command to Carey, accepted service to testify that day, after a threat by Attorney Tuttle to force the company to produce him.

First Officer Frank William Johnson, described by one reporter at the Tuttle hearings as "a wind-bitten little mariner," was brought to the federal building by Lamport & Holt officials at the request of District Attorney Tuttle, after agents of the department of justice had looked for him without success for two days to serve him with a subpoena to appear in court.

Johnson said he had been appointed second-in-command just before the *Vestris* sailed. He agreed with Tuttle that part of a first officer's duty was to see to the proper securing of coal and cargo ports before the ship left the harbor. He stated that there were two coal ports, each one three feet square and located about five feet above the nominal waterline. Each one was supposed to be fastened down by bolts from the outside of the ship.

"What did you do about closing the coal ports?" Tuttle asked.

"I told the ship's carpenter to close them," Johnson replied.

"Did you inspect personally to see if the work was well done?"

"No."

Tuttle then got Johnson to admit he had never inspected those ports "because there has never been any trouble with them."

"There was trouble enough with them on that trip," Tuttle replied. Then he asked if it was not true that gaskets, which were intended to make the ports watertight when they were bolted down, were missing.

Johnson said he knew of no places for any such gaskets in the ports but ended up admitting he was not sure whether there ever had been any gaskets.

Johnson said they first discovered that something was wrong Sunday night at seven o'clock. Tuttle reminded him that two wireless operators from the ship had testified to there having been a definite list as early as midnight Saturday.

"They were wrong," said Johnson.

Tuttle pointed out that in Capt. Carey's last wireless he stated that the ship had been hove to since Sunday noon.

"That was because of bad weather," Johnson said.

"But in that same message the captain described the weather as only moderately rough."

"It was more than moderately rough," Johnson insisted.

"You contradict your captain?"

"Yes."

Johnson said that Sunday evening the ship lurched heavily to starboard. The chief engineer told him that some cargo had broken through a bulkhead into the sailors' quarters on the starboard side. He inspected and found that a half door—a hinged opening in the side of the ship slightly larger than the coal ports and about the same distance from the water—near the bulkhead was leaking.

He said that on one side of the door, about as much water was coming in as is carried in a two-inch pipe. He had crewmen try to tighten up the bolts, which on this door were on the inside, but the door seemed to be sprung and the leak could not be stopped.

The cargo that went through the bulkhead weighed about ten tons, Johnson said, but the shifting of this much weight would not cause an appreciable list.

Johnson agreed with Tuttle that water entering the ship was responsible for the list, which increased during Sunday night to about thirty degrees.

"Where was the water coming in, besides that two-inch stream around the half door?" Tuttle asked.

"That's just what we tried to find out," said Johnson. He said that in his opinion "something had sprung" in the side of the ship and told of hearing water running in between decks where no one could get at it. This space between two decks is called a 'tween deck. He didn't know if the water was coming in faster than the pumps could handle it, and he didn't know what the capacity of the pumps was.

"You were the first officer on the ship," Tuttle said. "Have you no explanation as to what made it sink?"

"No."

"What time was it when you heard water between decks?"

"I'm vague about that."

"Oh," said Tuttle, "are you going to be vague about time too?"

Tuttle was undoubtedly referring to the failure of the radio operators to recall the timing of the CQ and SOS messages.

Tuttle asked Johnson, "When did you begin to think the bailing of the ship with buckets was useless?" Johnson replied, "I thought there was a chance all along. In my opinion the chances of sinking did not become greater than those of staying afloat until noon Monday." This was about two hours before the ship sank.

When Tuttle asked Johnson whether the captain sent any messages to the line officers or to other ships before or after the SOS, Johnson said, "I don't know."

When told to launch the lifeboats, the first officer assigned members of the crew to the work. Johnson said, "It was difficult to keep one's feet, and when I went along the port side I had to hang on to one or two of the passengers, as I could not get a grip on the rails." So bad were the conditions, he said, that only six of the fourteen boats were launched when the *Vestris* sank.[15]

At the British Board of Trade inquiry, Johnson said water that got into the 'tween deck and couldn't get out "capsized the ship."

A strong circumstantial case against Lamport & Holt

We back up now to the time when the *American Shipper* steamed into the port at New York on Wednesday, November 14, 1928. Standing on the deck was Captain Frederik Sorensen, a second-class passenger on the *Vestris*. He made a scathing statement to several reporters.

Sorensen asserted that the loss of the *Vestris* was caused by the procrastination of the captain, particularly regarding the late sending of an SOS; equipment that was in poor condition, especially the lifeboats and their lowering devices—he stated that one of the lifeboats was "rotten"; and confusion and lack of direction among the officers and crew of the ship.

All of these factors, Sorensen stated, were responsible for the loss of most of the women and all of the children who were aboard the *Vestris* when she foundered.

This statement was published in several New York newspapers, including the *Times* and the *New York World Telegram*. These versions of Sorensen's statement were in essential agreement, differing only in slight variations of wording, which served to show that they had not been copied from a common source

In addition to Sorensen's caustic remarks, some of the crew members of the *Vestris* were also critical of the way the disaster had been handled.

Three days later, on November 17, 1928, Captain Sorensen made a statement on the radio station WLTH (New York) in which he declared that he had been "shamefully misquoted," especially in his remarks about Captain Carey. He further asserted that a signed article, allegedly written by him, was printed entirely without his authority in a newspaper Thursday morning, saying:

I made none of the accusations against Captain Carey that were published. The only thing I said was that I thought he made a mistake in not calling for a ship to stand by and I suppose he realized that himself when it was too late.

Captain Carey knew if he was saved that he would be tried for criminal negligence and therefore sacrificed his life. In my opinion, he was a brave man. When he saw the ship was doomed, he followed the old traditions of the sea and went down with his ship. I think anybody with a heart would forgive a man's mistake, which he willingly paid for with his own life.

I am supposed to have said the crew were cowards, which is another lie. I did say that the crew of lifeboat number 1 were cowards or even worse when they saw our boat, number 8, sinking and refused to pick up the women and children despite their pleading. The majority of the officers and crew in my opinion were brave and true seamen and did wonderfully well under difficult circumstances.[16]

Obviously, this was a marked change from Sorensen's previous statement. An article in the British newspaper *Liverpool Post* told how *New York World Telegram* reporter James Duffy testified before the Hoover inquiry of the Steamboat Inspection Service. The day before Captain Sorensen had called Duffy a liar. Duffy, however, testified that his original report was correct. He read to the panel the original notes from his interview with Captain Sorensen and was emphatic that Sorensen said that a lifeboat of the *Vestris* was "rotten." He also testified that his story as given in the newspaper was printed under his own name, and he stated exactly how he had obtained the interview and what Captain Sorensen had said.[17]

In yet another interview, obtained by the *New York Times*, Fred W. Puppe, a passenger on the *Vestris*, said that Captain Sorensen described the crew of the Lamport & Holt liner as "murderers," and despite Sorensen's later retraction of this statement, Puppe insisted that the captain had been correctly quoted on his arrival here. Puppe, who was aboard the *American Shipper* when Captain Sorensen attempted to organize the survivors so they could press legal claims against Lamport & Holt, said that Sorensen was aggressive in his condemnation of the officers and crew of the *Vestris* and the condition of the ship and the lifeboats.[18]

It appears obvious from these later statements that Sorensen was not misquoted in his original remarks castigating the ship and its crew as "murderers." What would account for this remarkable shift in the captain's position? The answer is in the testimony given in the Tuttle hearing by Frank W. Johnson, the chief officer on the *Vestris*. Attorney Tuttle wrested from Johnson that he had been in contact with two Lamport & Holt lawyers, a Captain William Heasley, who was the shore captain (assistant marine superintendent), and "a man named Regan" who was the wharfinger[†] at Hoboken.[19]

The others, including no doubt Captain Sorensen, also had in all likelihood been in contact with the same Lamport & Holt officials. We suspect that they made Captain Sorensen an offer he couldn't refuse, something like "recant your statement or you will never be a ship's master again."

The entire transcript of the hearings conducted by Attorney Tuttle is rife with exchanges indicating that Tuttle did not believe some witnesses. He repeatedly asked why they had changed their stories and implied that he would bring perjury charges against any person whom he could prove had been lying on the witness stand.

A British Board of Trade inquiry in 1929 makes things clearer

This testimony is by Frank W. Johnson, chief officer on the *Vestris*, who was examined by attorneys E. A. Digby, Scanlon, and Stilwell.

[†] A wharfinger is the person in charge of a wharf.

§ Mr. Digby: "Did Mr. Anderson make any communication about being careful what draft you entered in the logbook?"

"Yes. He said, 'It is a winter voyage you know.'"

"Did you understand by that that you must not show a vessel loaded below her winter marks?"

"That is what I understood."

"We know that she did leave below her winter marks, and that draft was reported to you by the second officer?"

"Yes."

"Later, was there some conversation between you and the second officer about the draft?"

"No, I had no [further] conversation with him."

"Was there some comment about the draft?"

"Only by the captain."

"Will you tell us about that?"

"I went round to see everything was all right, and I went to the chart room and told the captain everything was all right. He spoke to me then about being careful what was put in the logbook about the draft."

"What did you understand by that?"

"That I had not to put in the logbook that she exceeded her winter draft."

Mr. Digby: "When you were rescued and got back, did you go to the Holly Hotel in New York?"

Mr. Johnson: "Yes."

"On the day you arrived were you visited by American lawyers acting in the interest of the owners?"

"Yes."

"Did you give them the information they wanted?"

"No, I was suspicious about them and did not say much."

"Was Mr. Regan (the wharfinger) at the hotel?"

"Yes."

"In the course of conversation was there any mention about the drafts of the *Vestris*?"

"Yes."

"Mr. Regan has told us he could not remember anything, and Captain Heasley said he began to have a glimmering about it. Was Mr. Regan present throughout the conversation?"

"Yes."

"Will you tell us in your own words what the conversation was?"

"Well, first, I think Captain Heasley asked me: 'What draft have you got?' I said: '26 ft. 6 in. and 27 ft. 11 in.' He said: 'Good God, no!' I said: 'What have you got?' He said: '26 ft. 2 in. and 27 ft 3 in.' And I said: 'That's good enough for me, then.'"

"The draft you gave us was what you had been told was the observed draft when the vessel left the pier?"

"Yes."

"What did you understand by the draft he gave you?"

Mr. Johnson (after a pause): "I knew there was going to be trouble about the draft. I knew that from the beginning."

"What draft did you understand he gave you?"

"The draft on leaving on her winter marks."

"As a matter of fact, I don't think you were asked at the American inquiry what the draft was?"

"No."

"Have you ever heard it suggested that documents relating to the *Vestris* have been destroyed in this case?"

"Yes."

"Will you say how and where you heard it?"

"I heard it on the dockhead at Hoboken. I went into Mr. Regan's office, and Mr. Regan left me to go out on the dock. While I was there the dockmaster came in. I have since heard his name was Lloyd. I did not know it at the time. He said to me, 'Oh, there was great excitement when the news came through about the *Vestris* going down.' He said, 'Heasley came into my office like a madman and tore up papers belonging to the *Vestris*.'"

Attorney Scanlan takes over the questioning: "You said that you knew at noon on Sunday that the vessel could not keep her course. Have you any explanation to give about that?"

"There was a rough sea at the time."

"Did it occur to you then that that was a matter which would justify the officers charged with navigating the ship in making some appeal for help?"

"No."

"When did it occur to you?"

"Not until ten o'clock on Monday morning, when I asked the captain if he had sent out an SOS and he said he had."

"Did it appeal to you as a man anxious for the safety of the ship that the SOS should have been sent out earlier?"

"No."

Attorney Stilwell: "You have given three reasons: bad weather, a tender ship, and the entering of water, as to why the ship foundered. Have you ever been inclined to add a fourth, namely deep loading? [After slight pause] Do you add it as a fourth reason now?"

"Yes."[20]

The hearing was then adjourned for the day.

The hearing resumes the next day

§ Frank Johnson was recalled and further questioned by Mr. Langton. He said that when longer hooks for the lowering gear of boats were introduced two or three years ago, they were not made applicable to old boats. Questioned about the entrance of water into the ship, Mr. Johnson said water was not coming through the half-doors sufficiently to account for the quantity in the ship, nor were the booby hatches forward allowing any substantial amount of water to enter.

The water that blew the hatch in the small bunker trimming hold must have come from below.

Langton: "When you got ashore, did you go to [a] hotel?"

"Yes."

"And you told us you were in communication with the lawyers; why were you suspicious of them?"

"I did not know who they were."

"And whilst you were in New York did you have any legal assistance?"

"Later on."

"Did someone come from the Navigators and General Insurance Company?"

"One of the managers, Mr. Coombs."

"From that time on did you always go to the solicitors for the Navigators and General?"

"No, it was not until towards the end of my stay in New York that I went to them."

The witness added that he could not get a ticket home from the Lamport & Holt office on Broadway; they said they had not got one for him, or at any rate, they had not got it ready; they might have been negotiating for it.

"How did you, in fact, get over here?"

"Mr. Coombs paid my fare."

"Has Mr. Coombs been here at this inquiry?"

"Yes."

Attorney Langton proceeded to question the witness about the statement he made yesterday: that Mr. Lloyd, harbor master at Hoboken, told him Captain Heasley tore up some of the *Vestris*'s papers after the disaster.

"You did not ask what papers?"

"No, I didn't know he had any papers. That's all he said to me."

"Have you been present during the whole of the inquiry?"

"Yes."

"Have you heard that there were no papers about the *Vestris* in Mr. Regan's office?"

"Yes."

"This is mere gossip. You are repeating this as something somebody said to you?"

"Of course he said it to me."

"Are you sure he did?"

"Yes."

"Upon that you constructed this charge against Captain Heasley of destroying documents?"

"I have not constructed any charge at all that I know of."

"So far as you know, is there anything else upon which this charge is built?"

"I don't know of anything."

"Without probing the matter anymore you instructed solicitors or counsel to put forward this charge?"

"No, I have not instructed anyone, but decided I would keep nothing back."

"You appreciate, of course, that it reflects very seriously on Captain Heasley's character?"

"The whole thing is serious from beginning to end."

"Having heard the evidence that there were no documents, has it any other relevance to this inquiry?"

"I don't know."

"Can you suggest any reason why it should be brought up except to blacken Captain Heasley's character?"

"I don't know anything at all about it; I don't understand these legal matters."

"Let us pass on to the other charges that you have made, and let us first of all try to get the facts about the logs. Do I understand that you had made no entry in your log about the draft?"

"I had not started to write up the log."

"The scrap log had been entered up, had it not?"

"That was kept by one of the other officers."

"Don't you know whether it had been entered up?"

"I don't know whether the draft had been entered up. The other particulars would be entered."

"Haven't you had cause to inquire whether it was entered up?"

"No."

"Whose business is it to enter up the scrap log?"

"Mr. Welland's."

"If Mr. Welland says the entry he made was 27 feet 11 inches aft and 26 feet 6 inches forward, that would be right, would it not?"

"Yes, from the information I got from Mr. Watson."

"When you come to enter your log, where do you take the draft from?"

"It was my intention to talk to the captain of the ship about it and see what draft we should put in the log book."

"Up to that time you had had one conversation with Mr. Anderson and one with Captain Carey about the draft?"

"Yes."

"When you had that conversation with Mr. Anderson, did you make any protest about what you suggest he meant you to do?"

"No."

"Did you know that what he meant you to do was a criminal offense?"

"Yes."

"Do you suggest to this court that you were ready to commit a criminal offense without protest?"

"Yes."

". . . Are you seriously suggesting that Captain Carey meant you to put a false entry in the official log?"

"Yes."

"Captain Carey is not here to defend himself."

"I know, and that is what you are taking advantage of in my opinion. I don't like saying these things."[21]

But say them he did, and after that the fat was in the fire. Lamport & Holt was done for.

Other contributing factors

In addition to the factors discussed previously, there are several other things that contributed to this disaster. Most significant of these was that the *Vestris* was overloaded when she left port. There was some controversy over this, but Judge Goddard in his 1932 court opinion stated, "I find the mean [draft] of the *Vestris* for this voyage was 26 feet 11½ inches." [See Appendix A for complete decision.]

He then concluded by writing:

In 1912, upon leaving her builders in Belfast, load line or "Plimsoll" marks were affixed to the *Vestris* in accordance with the British Merchant Shipping Acts and the draft standards of the British Board of Trade. Under this standard she was assigned a salt water draft of 26 feet 3¼ inches for winter and 26 feet 9¾ inches for summer voyages. So that the *Vestris* sailed on this winter voyage loaded 8¼ inches below her assigned load line marks for winter voyages, and even below her summer marks by 1¼ inches.[22]

This was a critical finding since it meant that the potentially leaky ports were brought even closer to the level of the sea than they should have been.

At the Board of Trade inquiry, First Officer Leslie Watson made this point quite clearly. When questioned by Attorney Scanlon about what he knew about the overloading, Watson testified:

§ I had an idea that Mr. Coombs knew something was wrong somewhere, and the chief officer [Johnson] was very nearly crazy about the whole affair; we stayed up all night talking about it. . . .

"Well, I am going to tell him that the ship was overloaded," I told him when he came the next morning. I was fed up. I knew that when I came over here I would have to tell the Board of Trade, and so I said to Mr. Coombs: "Here's the true draft; you can have it." I gave it to him and subsequently made an affidavit declaiming the true draft.

Asked by Attorney Scanlon the next day if he had formed any theory as to why the ship sank, Watson answered, "Yes. A combination of circumstances. It was not just one thing."

"Is one of the circumstances that she was overladen?" asked Scanlon.

"Yes," replied Watson. "That has got a lot to do with it."[23]

THERE WERE ALSO A NUMBER OF ALLEGATIONS made that Captain Carey was in contact with Lamport & Holt officials during the day on Sunday and was told that he should try to avoid abandoning ship at all costs. These allegations could not be proved, but it appears that they were based on prior instructions made by the owners to the masters of their vessels.

The British Board of Trade's inquiry found written instructions stating, "In the event of a serious disaster happening to one of the vessels of this line while at sea, the master must in the first instance calmly consider the actual amount of peril there may be to the lives of those under his charge, and then judge if he will be justified in fighting his own way unaided to the nearest port. His being able to succeed in this will always be considered as a matter of high recommendation for him as master."[24] This was no doubt an effort to avoid salvage claims[†] and one reason for the delay in sending a distress call.

[†] The law of salvage is a concept in maritime law which states that a person who recovers another person's ship or cargo after peril or loss at sea is entitled to a reward commensurate with the value of the property so saved. The concept has its origins in antiquity, with the basis that a person would be putting himself and his own vessel at risk to recover another and thus should be appropriately rewarded. Salvage law has been recognized for

Bodies of eleven dead arrive at Staten Island

Dead bodies wrapped in canvas on deck of the USS *Shaw*. Photo taken November 14, 1928, by *Daily News* photographer Ed Jackson.

Vestris survivors on the deck of the *American Shipper*

The man holding the cat is steward Alfred Dineley, who was picked up by the *Shipper*. Photo taken November 14, 1928, by an unnamed photographer.

Vestris survivors on the deck of the German ship *Berlin*

The ship is identified by her life preserver ring, held by a man seated in front. Photo taken November 14, 1928, by *Daily News* photographer Bob Costa.

Surviving crew members of the *Vestris* being paid

Officials of Lamport & Holt paying surviving crew members while in New York. Photo taken November 15, 1928, by *Daily News* photographer Hank Olen.

Captain Carey was evidently hoping a northbound sister ship would arrive in time to assist. That ship was the *Voltaire*, but she allegedly had propeller problems. The *Vestris* radioed the *Voltaire* at least three times late Sunday and Monday morning, each time saying it had "nothing to communicate." The *Vestris* was, however, in contact with radio station WSC at Tuckerton, New Jersey, on Monday, until 1:35 P.M. Less than one hour later, the *Vestris* turned over and sank.

The lawyers arrive

The *Vestris* had not been in her watery grave for even two weeks when the first lawsuit was filed. [§] On November 24, 1928, Orrin S. Stevens of Boston, a survivor of the *Vestris*, filed suit in federal court for $52,000 in damages against the owners and operators of the ship—$50,000 for the death of his wife and $2,000 for the loss of his luggage. He alleged that the *Vestris* was not seaworthy; that she was loaded improperly, causing a list; that her officers and crew were

---

centuries in such documents as the edicts of Rhodes and the Roman Digest of Justinian. It is still a nearly universally recognized right, though conditions for awards of salvage vary from country to country.

The right to be rewarded for salvage at sea under common law is based both on equitable principles and public policy and is not contractual in origin. Historically, salvage is a right in law, when a person, acting as a volunteer (that is, without any pre-existing contractual or other legal duty so to act) preserves or contributes so to preserving at sea any vessel, cargo, freight, or other recognized subject of salvage from danger.

The law seeks to do what is fair to both the property owners and the salvors. The right to salvage may not necessarily arise out of an actual contract but is a legal liability arising out of the fact that property has been recovered. The property owner who had benefit of the salvor's efforts must make remuneration, regardless of whether he had formed a contract or not. The assumption here is that when faced with the loss of his vessel and cargo, a reasonable, prudent owner would have accepted salvage terms offered, even if time did not permit such negotiations.   —*Wikipedia*

incompetent; and that she was not supplied with adequate, working lifesaving equipment.

Stevens, a resident of Buenos Aires followed up the filing of his suit with a plan to organize the rescued passengers into a committee for joint legal action [what we now call a "class action"].

Mr. Stevens also filed for libel against the Liverpool, Brazil, and River Plate Steam Navigation Company, owners of the *Vestris*, and Lamport & Holt Ltd., the operator. He also obtained from the judge in his case an order for subpoenas directing 123 members of the ship's crew to appear for examination.[25]

Lawsuits were eventually brought against Lamport & Holt by 60 claimants totaling some five million dollars.

An interesting aside is that the sinking of the *Vestris* was covered by Associated Press reporter Lorena Hickok. Her story became the first article to appear in the *New York Times* under a woman's byline. She went on to become one of Eleanor Roosevelt's closest confidants and a lead investigator for Federal Emergency Relief Administration head Harry Hopkins from 1933 to 1936.

Fallout: SOLAS 1929

The *Vestris* disaster resulted in far-reaching consequences. Captain Overstreet, the commander of the battleship *Wyoming*, had been rather critical of the life jackets used on the *Vestris*. It seems that a number of bodies had been found floating face down in their life jackets. Captain Overstreet recommended replacing the cork life vests with kapok life jackets, which would hold the head of an unconscious victim out of the water so they would not drown.

The 1929 International Convention for the Safety of Life at Sea (SOLAS) made several changes in existing rules governing passenger ships. Some of these new rules covered problems with lifeboats and others dealt with the efficiency of life jackets. Here is some text taken directly from the official report of the 1929 SOLAS Convention.

ARTICLE 13

Lifeboats and Buoyant Apparatus

The general principles governing the provision of lifeboats and buoyant apparatus in a ship to which this Chapter applies are that they shall be readily available in case of emergency and shall be adequate.

1. To be readily available, the lifeboats and buoyant apparatus must comply with the following conditions:—

(a) They must be capable of being got into the water safely and rapidly even under unfavorable conditions of list and trim.

(b) It must be possible to embark the passengers in the boats rapidly and in good order.

(c) The arrangement of each boat and article of buoyant apparatus must be such that it will not interfere with the operation of other boats and buoyant apparatus.[26]

This article has been widely interpreted as a reaction to the details of the *Vestris* disaster. Life jackets were redesigned as a result.

Eventual legal disposition

The legal issues proceeding from the sinking of the *Vestris* wound their way through the courts for fully four years. In an attempt to avoid going bankrupt from the many lawsuits filed against them, the Lamport & Holt firm filed a petition with the US District Court for the Southern District of New York asking that their liability in the matter be limited.

In his Opinion on the Merits, dated May 24, 1932 (see Appendix A for the complete text of this opinion), Judge Goddard stated, "Numerous suits having been brought against them, the petitioners in this proceeding seek to obtain an adjudication as to liability, and to limit it to their interest in the vessel and her pending freight should the court find liability."[27] He found that Lamport & Holt could not claim limited liability in the sinking of the *Vestris* because they had been negligent in too many ways to allow such a finding. He therefore found that they would be liable in unlimited amounts.[28]

Lamport & Holt could not satisfy the potential judgment against them, since their business had understandably dried up following the *Vestris* disaster, and the ensuing Great Depression had further distressed everyone in the shipping business. They therefore made a counteroffer to the claimants. This is revealed by a letter from the offices of the attorneys handling the case to their clients.

§ New York, N. Y.

November 23, 1932.

S.S. *Vestris*

Dear Sirs:

A meeting of proctors for claimant (or claimants) will be held at the office of Messrs. Bighain, Englar, Jones & Houston, 9th Floor, No. 64 Wall Street, on Tuesday, November 29th, at 10:30 A.M. to consider the following recommendation of settlement:

Sir William McLintock, G.B.E., C.V.O., the Receiver for the Debenture holders in Lamport & Holt, Limited, has been in negotiation with your Committee for some time with a view to a settlement of the litigation in this country resulting from the loss of the *Vestris*. In this connection he has put before your Committee the financial position of the Liverpool, Brazil and River Plate Steam Navigation Company, Limited, owner of the *Vestris*, and Lamport & Holt, Limited, operator of the vessel. So far as Lamport & Holt, Limited, is concerned, your Committee has been satisfied that there are existing liens against the property of this Company entitled to priority over the *Vestris* claims for an amount exceeding the value of the assets of the Company. Your Committee is convinced that in no event can the

Vestris claimants make any recovery from that Company. The Receiver has also put before us the accounts of the Liverpool, Brazil and River Plate Steam Navigation Company, Limited, and has fully discussed with us the priorities of various classes of claims, the value of the fleet owned by that Company, and heavy depreciation of the Company's assets by reason of the present world depression in the shipping industry.

The Liverpool Brazil Co. is prepared to pay £110,000 in settlement of the American litigation. This amount is to be distributed among all the claimants pro rata in accordance with the amount of their claims liquidated in time amounts that may be fixed by the Committee on the basis of the questionnaires supplied by the claimants. Your Committee is satisfied after these discussions that this amount is substantially more than can be recovered in the event the litigation is continued and the Liverpool, Brazil and River Plate Steam Navigation Company, Limited, is forced to realize its assets. Your Committee appreciates that the amount of the proposed settlement is very small in comparison with the damages which can be proven in the proceeding; but your Committee has made every effort to secure a better settlement and is clearly of the opinion that this is the best figure that can be obtained.

If the settlement is not accepted, the litigation in this country will continue, the ship owners' appeal will be prosecuted, very substantial expense will have to be incurred, and a final adjudication can hardly be hoped for in this country in less than one or two years. This would have to be followed by legal proceedings in England, the length of which it is difficult to determine, but English counsel have intimated that it might take a further two years. Under present conditions it will be realized that the future of the shipping industry and particularly of companies engaged therein is one of great uncertainty. Your

Committee also recognizes that a realization on the ships belonging to the Liverpool Brazil Company in these depressed times would be a matter of much difficulty.

While your Committee is disappointed that the amount to be received is not larger, your Committee is unanimously of the opinion that it is in the interest of all the claimants to approve the settlement, and recommends its acceptance.

The detailed terms of the settlement are given in the enclosed memorandum, and your Committee requests you to sign one copy of this memorandum and return it to the Committee as promptly as possible so that the settlement may be carried out.

If you have not heretofore supplied the Committee with a questionnaire as a basis for determining the amount of your claim, or if you desire the Committee to consider any further evidence or information in connection with the claim, please send it to the Committee as promptly as possible as it is essential that the Committee should conclude its liquidation at the earliest possible moment if this settlement is to be carried through at all.

> Very truly yours,
> OSCAR R. HOUSTON, Chairman,
> GEORGE WHITEFIELD BETTS, JR.,
> CHARLES R. HICKOX,
> RUSH TAGGART,
> Committee.[29]

At the time this was written, the British Pound Sterling was equal to five US dollars, so the amount of the offer was about $550,000. This was roughly eleven cents on the dollar of the five million in damages sought by the claimants. Each claimant received approximately $9,170, assuming that no attorney's fees were taken out of the amount of the settlement. Considering the state of the economy in 1932, this offer was probably the best they could have expected.

To put the amount of this settlement into perspective, $9,170 in 1932 was equivalent to about $164,000 in 2015 dollars.

Press evaluation of the *Vestris* disaster

The *Augusta Chronicle* ran the following op-ed piece under the title "The Truth About the *Vestris*" on May 9, 1929:

> Captain Carey of the ill-fated *Vestris* died with his ship and 111 lives were lost. An American inquiry was conducted which did not reveal, apparently, the true cause of the disaster, which according to Frank W. Johnson, chief officer of the *Vestris*, was overloading. This was concealed at the American inquiry into the disaster because it was not desired that the American people would know the real cause of the disaster. This leads the *Chicago Tribune* to remark:[30]

> This testimony is almost incredible, but interested British cross-examination left it standing and unimpeached. Captain Carey paid with his life. There were 111 lives lost when the *Vestris* sank and in the light of the chief officer's testimony they were as good as murdered to show a profit on the trip.

> What may never be known is the influence brought to bear on a ship master, now dead, to disregard safety and load in the cargo. That evidence might be developed in London from the owners. It would fit in to explain also why the captain fought to avoid a salvage charge against the owners.

> What can be ascertained here in the United States is why the federal inspection permitted an overloaded and unseaworthy boat to sail. Profits may have impelled a British ship to take her passengers out to death, but what connivance or indifference, negligence or ignorance in an American port contributed to the disaster?[31]

The *Vestris* sailed from an American port. The overloading was done at an American port and as the *Tribune* well says, "What connivance, ignorance, or neglect in an American port contributed to the disaster?" It would appear that our ship inspection is but a mockery, if such could happen and the inspectors not know it. The lesson is for it not to happen again by prompt punishment of those responsible.[30]

It certainly sounds as though the US media were up in arms about this sorry incident. One can only wonder if any punishment was ever meted out to those responsible for the tragedy. We could find no record of this having happened. We suspect that nothing was ever done about those who were ultimately to blame for this terrible loss of lives, and that it was simply swept under the rug.

～ ❀ ～

Vestris Timeline

Date	Event
Nov. 5–7	*Vestris* in dry dock for inspection.
Nov. 9	*Vestris* has collision with another ship on its way to Hoboken pier.

Saturday, November 10, 1928

Time	Event
3:45 P.M.	*Vestris* sets sail from Hoboken pier.
4:45 P.M.	Water is coming through the ash ejector.
About 8 P.M.	List of ship becomes noticeable.

Sunday, November 11, 1928

12:00 A.M.	Water seen coming through the half door (aka working door or coal port).
2:00 A.M.	Second Officer Watson notices list at five degrees to starboard.
4:00 A.M.	Seaman Fred Gill sees water pouring through coaling doors on port side. List continues at a greater angle. Bailing begins.
5 A.M. – 7 A.M.	Captain notes situation getting serious.
6:00 A.M.	List makes moving around staterooms difficult. Five passengers show for breakfast. Others are seasick.
7:00 A.M.	Crew hear water running underneath coal.
8:00 A.M.	Moderate gale. Ship rolls eight to nine degrees starboard.
9:00 A.M.	Crew discovers leak in ash ejector starboard.
10:00 A.M.	Leak is discovered in working door portside. Leak in lavatory also found.

Sunday, November 11, 1928, continued	
Time	**Event**
11:00 A.M.	*Vestris* is hove to.
Noon	Leak in ash ejector and lavatory plugged. Storm increases and ship remains hove to. List so pronounced passengers have to hold plates on dining table.
AFTERNOON	Several staterooms are flooded. Furniture in the ship's salons is fastened down because of storm.
4:00 P.M.	Weather is characterized as a gale. Leak in coal bunker starboard. Three leaks altogether: two in the coal bunker—one starboard and one portside—and one in a working door portside.
7:30 P.M.	Wave causes a heavy lurch on ship; cargo shifts and breaks through bulkhead. List becomes permanent between five and fifteen degrees.
8:00 P.M.	Storm is severe.
9:30 P.M.	Sea calms down somewhat.
Monday, November 12, 1928	
3:00 A.M.	All hands called to start bailing water.
MIDNIGHT–4:00 A.M.	List at twenty degrees. Starboard boiler is closed down. Bucket brigade bailing water.
6:00 A.M.	Crew throws some cargo overboard.
7:00 A.M.	Food is all spoiled. List now twenty to thirty degrees.
8:00 A.M.	Bucket brigade gives up bailing as useless.

Monday, November 12, 1928, continued	
Time	**Event**
8:37 A.M.	*Vestris* radios CQ distress message.
9:00 A.M.	Women and children in third-class rooms ordered to first-class deck. Captain Carey orders women and children on shelter deck in anticipation of loading the lifeboats. Pastry chef Fred Hanson begins taking pictures with his box camera.
9:56 A.M.	First SOS is radioed by *Vestris*.
10:30 A.M.	Women and children told to put on life belts.
11:00 A.M.	Captain Carey estimates the *Vestris* is listing thirty-two degrees. Measurements of photos taken at noon show a twenty-nine degree list.
11:06 A.M.	*American Shipper* receives news of SOS from *Vestris*.
11:30 A.M.	Begin preparing lifeboats for launching.
11:40 A.M.	Captain Carey orders lifeboat launching to begin.
Noon	Start loading women and children in lifeboats.
1:00 P.M.	Engineers leave engine room. Engine room, stokehold, and bunkers all flooded.
1:30 P.M.	No lifeboats have reached the water yet.
1:30 – 2:30 P.M.	Only nine of fourteen lifeboats are launched. One later sinks.
2:36 P.M.	The *Vestris* sinks.
7:30 P.M.	*American Shipper* arrives at position given in SOS.

Monday, November 12, 1928, continued	
Time	**Event**
8:00 P.M.	Passenger William W. Davies in lifeboat No. 1 sees sweeping searchlight. People in lifeboat No. 3 also see searchlights.
9:00 P.M.	Those in lifeboat No. 10 see searchlights.
11:08 P.M. until 2:00 A.M.	*American Shipper* searches for wreckage.
Tuesday, November 13, 1928	
3:00 A.M.	Searchlights again spotted by people in lifeboat No. 3.
3:40 A.M.	*American Shipper* sights red flare.
4:00 A.M.	Those in lifeboat No. 10 see *American Shipper* searchlight.
4:05 A.M.	*American Shipper* picks up lifeboat No. 5.
4:30 A.M.	*American Shipper* picks up lifeboat No. 1. *Myriam* picks up lifeboat No. 11.
5:15 A.M.	*American Shipper* picks up lifeboat No. 3 with forty-five persons aboard.
5:40 A.M.	*Myriam* picks up lifeboat No. 7.
DAWN	About 6:45 A.M. *Berlin* picks up lifeboat No. 13 with thirty-two or thirty-three persons on board.
7:00 A.M.	*American Shipper* picks up lifeboat No. 10.
7:06 A.M.	Sun rises over area where lifeboats are located.

Tuesday, November 13, 1928, continued	
Time	**Event**
7:30 A.M.	*American Shipper* picks up lifeboat No. 14.
10:00 A.M.	*Berlin* picks up Carl Schmidt from water. *Wyoming* picks up Marie Ulrich, Gerald Burton, Marion Batten, and John Morris from the water; they were survivors of capsized lifeboat No. 8. USS *Wyoming* also picks up four others from two improvised rafts.
11:00 A.M.	Paul Dana and Clara Ball sight the *Wyoming* and start swimming for it.
NOON	Paul Dana and Clara Ball sight the *American Shipper*. Shortly later the *American Shipper* picks them up.

The Aftermath	
Date	**Event**
Nov. 15, 1928 to Dec. 6, 1928	Investigation held at the Federal Building in New York City under US Commissioner Francis O'Neill, led by District Attorney Charles H. Tuttle. Finds the captain and crew of the *Vestris* culpable.
Nov. 20, 1928 to Dec. 14, 1928	A hearing is held in the Custom House in New York under Dickerson N. Hoover, Supervising Inspector General of the Steamboat Inspection Service of the Department of Commerce. Concludes inspections not to blame.
Nov. 25, 1928	The first lawsuit is filed against Lamport & Holt on behalf of survivor Orrin S. Stevens. Eventually some sixty survivors join in a class action suit.

The Aftermath, continued	
Date	**Event**
April 22, 1929 to July 31, 1929	British Board of Trade inquiry into the *Vestris* held in London. Inquiry concludes July 5, 1929, the longest in British history. Judgment rendered July 31 finds that the principal causes of the disaster were overloading of the ship and its relatively poor condition, which resulted in numerous problems when a major North Atlantic storm came up.
April 15, 1931	Lawsuit brought by Lamport & Holt to limit their liability in the *Vestris* disaster begins in New York Admiralty Court; United States District Judge Henry W. Goddard presiding.
May 24, 1932	Judge Goddard delivers his opinion on the merits of the Lamport & Holt action, finding that they cannot claim limited liability for the wreck of the *Vestris*.
Nov. 23, 1932	Attorneys for the claimants against Lamport & Holt deliver the final offer of $550,000 for the loss of the *Vestris*, amounting to about $9,170 per claimant. This is only 11 cents on the dollar of the $5,000,000 asked for in the lawsuit.

*Appendix A: Judge Goddard's Opinion on the Merits
in the Petition of Lamport & Holt to Limit Liability*

(Reprinted from 1932 A. M. C.)

**In the United States District Court for the
Southern District of New York**

VESTRIS

DECISION ON THE MERITS

PETITION OF THE LIVERPOOL, BRAZIL & RIVER PLATE STEAM
NAVIGATION CO., LTD. AND LAMPORT & HOLT, LTD. FOR
LIMITATION OF LIABILITY, AS OWNERS OF THE S. S. *VESTRIS*.

OPINION OF GODDARD, DISTRICT JUDGE[†]

May 24, 1932.

BURLINGHAM, VEEDER, FEAREY, CLARK & HUPPER (VAN
VECHTEN VEEDER, CHAUNCEY I. CLARK, PAUL FEARSON
SHORTRIDGE and STANLEY R. WRIGHT, Advocates), *Proctors for
Petitioner.*

BIGHAM, ENGLAR, JONES & HOUSTON (D. ROGER ENGLAR, OS-
CAR R. HOUSTON EZRA G. BENEDICT Fox, W. J. NUNNALLY, JR.
and FRANCIS B. REED. Advocates), *Proctors for 27 death claims, 8
personal injury claims, 13 baggage claims, 550 cargo claims.*

[†] President Warren Harding in 1923 appointed Henry Warren Goddard as a
judge for the United States District Court for the Southern District of New
York. Goddard remained there as an active judge until 1954.

HUNT, HILL & BETTS (GEORGE WHITEFIELD BETTS, JR., GEORGE C. SPRAGUE. EDNA RAPALLO and JOSEPH A. MCDONALD, Advocates), *Proctors for Frederick F. Brown, etc.*

GEORGE Z. MEDALIE, US Attorney. (CHARLES E. WYTHE, Advocate), *Proctor for United States.*

HAIGHT, SMITH, GRIFFIN & DEMING (ARNOLD W. KNAUTH, Advocate), *Proctors for Elvira F. Rua, as admx. of Jose Gonsalves Rua, etc.*

KIBLIN, CAMPBELL, HICKOX, KEATING & MCGRANN (CHARLES R. HICKOX and CLEMENT C. RINEHART, Advocates), *Proctors for West India Oil Co., H. C. Johnston and E. Lehner.*

CARTER, LEDYARD & MILBURN (RUSH TAGGART, Advocate), *Proctors for C. P. Andrews & Co.* and *certain other cargo claimants.*

RAMSEY & MORGAN (MARK W. MACLAY and RALPH W. BROWN, Advocates). *Proctors for Marion C. Batten, ind. and as admx. of Norman Kirkpatrick Batten, deceased.*

CHOATE, LAROCQUE & MITCHELL, *Proctors for Mary L. Stone, as admx. of Charles I. W. Stone, deceased.*

LORD, DAY & LORD (F. J. CLARK, Advocate), *Proctors for Herman Hipp.*

NATHAN BURKAN (HERMAN FINKELSTEIN, Advocate), *Proctor for claimant, Anne DeVore.*

WISE, SHEPPARD & HOUGHTON (ARNOLD W. KNAUTH, Advocate), *Proctors for Otto Willi Ulrich and Maria Willi Ulrich, his wife.*

EDWARD J. LEON (SIDNEY M. OFFER, Advocate), *Proctor for Michael Coriarty.* [Ed. Note: Name spelled "Khoriaty" in passenger list.]

KIDDLE, MARGESON & HORNIDGE, *Proctors for G. H. Halpert.*

L. V. AXTELL, *Proctor for various death and injury claimants.*

Previous proceedings reported at 1931 A. M. C. 1553 (Preliminary opinion on three questions of law), and at 1931 A. M. C. 64, 755, 1914, and 1932 A. M. C. 608 (concerning security).

Henry W. GODDARD, District Judge:

This is a petition for limitation of liability filed in behalf of the owners of the steamship *Vestris*, which sailed from Hoboken, New Jersey, on the afternoon of November 10, 1928, on a voyage from New York to South American ports. On the 12th day of November, 1928, she foundered at sea with the loss of one hundred and [eleven] lives, the vessel, cargo and the effects of the passengers and crew. Numerous suits having been brought against them, the petitioners in this proceeding seek to obtain an adjudication as to liability, and to limit it to their interest in the vessel and her pending freight should the Court find liability.

At the time of her loss, the *Vestris* was on a voyage from New York to Barbados and South American ports with 325 persons aboard consisting of 128 passengers and a personnel of 197, including officers and crew; also cargo made up chiefly of motor parts and fruits. The estimates of the total weight of cargo, coal, water, stores, baggage, mails, kentledge,[†] and water in her ballast tanks when she sailed on this voyage from New York varied between 7370 and 7665 tons.

I. The *Vestris* was a British steel-passenger steamship of 10,494 tons gross and 6,622 tons net register with twin screws and was 496 feet long, 60 feet 6 inches beam. Her weight light was 7914 tons in addition to 1401.74 tons of refrigerating machinery and refrigerating insulation installed two years after she left her builders. The salt water draft assigned to the *Vestris* in 1912 (when she left her last British port) by Lloyd's for the British Board of Trade was 26 feet 9¼ inches for summer, and 26 feet 3¼ inches for winter voyages. Her salt water draft on this—a winter voyage—I find to have been 26 feet 11½ inches. She was built in 1912 at Belfast, Ireland, of the Isherwood type of construction; that is—with the longitudinal framing system. She was built to Lloyd's specifications and was classed "100 A 1" by Lloyd's Registry, which classification she retained after surveys made from time to time by its surveyors. Since 1921 the *Vestris* had been engaged in trade between New York and South American ports. Her British passenger certificate expired in 1922. But she was subject to the United States Steamboat Inspection Service which made yearly inspections of her and she held a United States Passenger Certificate and Certificate of Inspection

---

[†] Kentledge: permanent metal ballast. See "Obscure nautical terms" on page 226.

of boats and life-saving appliances. She carried fourteen clinker-built life boats which had a total authorized capacity of eight hundred persons. All the boats, with the exception of Numbers 13 and 14, which were extras, were equipped with "Martin" davits and "Mills" releasing gear.

She was a shelter deck vessel, having three complete decks extending from stem to stern, namely shelter deck, upper deck, and main deck; also a lower deck extending from her stem to the after end of No. 2 cargo hold forward. On the shelter deck was a cross alleyway extending the width of the ship. On the starboard side leading forward from the cross alleyway was the firemen's passage or alleyway. The forward part of the shelter deck was an open deck generally known as a "well deck." It is the open space between the forward end of the bridge house bulkhead and the island erection forward. It is 32 feet fore and aft at the sides and there is no bulwark but only an open iron rail. In the well deck were two booby hatches (closed companionways) for the use of the crew in going to and from their quarters to the stokehold; one on the starboard and one on the port side, each located about 18 feet from the ship's side and their after-sides being the forward thwart-ship bulkhead of the bridge house. They are constructed of steel and riveted to the deck. The openings of the booby hatches were 3 feet 6 inches high and 2 feet wide and faced outboard with 18 inch sills. They had double hinged wooden folding doors 2 to 2½ inches thick. The starboard booby hatch led to the firemen's alleyway. From the firemen's alleyway and the cross alleyway there were four large openings or doorways leading into the starboard shelter 'tween deck bunker. These openings were provided with loose plates to prevent coal from falling out, but did not stop water from going through. In the starboard shelter 'tween deck bunker there were two large and one small openings or hatches. There were covers for these hatches but they were not used and the hatches were left open so that the coal would run down into the lower bunkers as it was used. If water came through the booby hatch it had access through the firemen's alleyway and the cross alleyway and it would overflow into the starboard shelter 'tween deck bunker if it got above the coamings which were nine to twelve inches high. At each end of the cross alleyway is an opening on the side of the ship; these openings are six feet high and 4½ feet wide, known as "working doors" and are closed by doors known as "half doors"; these doors are divided into an upper and lower half. Before leaving New York these doors had been closed and white lead and spun yarn packed into the spaces between the doors and the sides of the ship. But it appears that the starboard half door "stays out a little bit"; that also on the port side the upper half door did not meet the lower one so that a space of about one-half inch was left. The bottoms of these openings were approximately 4 feet above the water line as the *Vestris* was loaded when she left New York on this voyage. Each degree she listed brought the starboard half doors about six inches nearer to the sea.

At 3:45 o'clock on the afternoon of Saturday, November 10, 1928, the *Vestris* left the pier at Hoboken for a voyage from New York to South American ports. At the time of her sailing the weather was fine and clear and according to the credible testimony she had no list. She continued under these conditions until soon after midnight when a slight list to starboard was noticed. By 5 A.M. she had a list of two to four degrees to starboard and two hours later the list had increased to not less than four or five degrees. Around midnight the wind had freshened and from then until about 7:30 the following morning there was a fresh northeasterly breeze. The wind at 7:30 was about 35 miles an hour. During the latter part of the night water began to come into the stokehold through the starboard ash ejector discharge and continued until about noon on Sunday when the leak was stopped. A considerable amount of water came in here and from other sources, for before noon the starboard bilges were full, and the water was up to the floor plates of the engine room which were about three feet above the tank tops. Water had also come through an open space in the port half doors into the cross alleyway, but these doors were re-caulked and made tight during the forenoon. Early Sunday morning water began coming in through the starboard booby hatch to the firemen's passage in the shelter deck and later the water from the firemen's passage was overflowing into the starboard coal bunkers. During Sunday evening the wooden doors of the starboard booby hatch were carried away and larger quantities of water came in through the opening and overflowed into the starboard bunkers. By noon, with the wind on her port quarter, she had a list of between 6 and 8 degrees and the wind and sea had risen; and at 12:20 P.M. she was hove to and allowed to lie in the trough of sea with the engines being operated intermittently at slow speed to keep her up to the wind. In the forenoon the caulking around the starboard half doors was washed out as the doors did not fit properly and water from here ran into that part of the ship and came out of the after end of the starboard side of the stokehold. Several unsuccessful efforts were made to put in new packing, but it would not stay as the doors were sprung. Beginning Sunday morning water entered the firemen's passage through the starboard booby hatch and finally overflowed into the starboard bunkers. At 2 o'clock Sunday afternoon, upon being informed by the Chief Engineer of the condition of the engine room and stokehold, the captain ordered starboard ballast tank No. 4 pumped out, and during the latter part of the afternoon No. 5 starboard ballast tank was pumped, but as these pumps had no wing suctions loose water was left in them. Early Monday morning starboard ballast tank No. 2 was pumped. The effect of pumping out these tanks was to increase the list instead of reducing it.

The weather and sea grew worse during the afternoon and at about 7:30 P. M. a heavy sea broke against her port side and she lurched heavily to starboard and some of her cargo in No. 1 hold of her upper deck shifted, breaking a temporary wooden partition. After the lurch she had a list to starboard variously estimated from ten to fifteen degrees. The list gradually increased until she capsized; at 4 A.M. Monday it was about twenty degrees; by noon it had increased to about thirty-five degrees. Just how much more she listed in the period between noon and the time she turned over is uncertain. The storm was most severe during Sunday evening and the wind reached about force 10 on Beaufort's scale (wind at 56 to 65 miles an hour).

Between 3 and 4 A.M. on Monday morning, the cover on the starboard hatch in the cross alleyway leading to the after cross bunker was blown up by the pressure of water in the bunkers. At 4 A. M. Monday the water had risen so high that the starboard boiler was closed down, and although her pumps, with the exception of the circulating pump of the engine, were in operation much of the time the water was gaining and a bucket gang was organized to bail water from the cross alleyway. This, however, was ineffective and was discontinued at 8 A. M. At 8:37 A.M. the alarm signal "C. Q." was sent out. At 9:56 A. M. the "S.O.S." signal was sent out. This was answered by fifty-eight ships and a number of shore stations. The "S.O.S." signal sent by the *Vestris* gave her position as Lat. 37° 35' N. and Long. 71° 81' [*sic*] W. which was incorrect by some 37 miles.† Sanderson & Sons, the New York agents for the

---

† There is no such thing as 81' in longitude. The original, printed court document has this, but it is obviously a typo. Information from other sources shows that the position as given in the SOS message was actually 37° 35' north, 71° 08' west.

owners, upon being informed of the S.O.S. call by the *Vestris* sent at 10:40 A.M. the following message by radio to her: "Wire us immediately your trouble." At 11 A.M. the captain of the *Vestris* replied: "Hove to from noon yesterday during night developed 32 degrees list starboard impossible proceed anywhere sea moderately rough." At 11:27 A.M. the representatives of the owners sent the wireless message to the *Vestris*: "United States destroyer *Davis* proceeding to your assistance," and they also immediately dispatched a large ocean-going tug with salvage equipment.

Shortly before 1 P.M. the water-tight bulkhead separating the engine room from the stokehold burst and soon after this the engineers left the engine room and went up on deck because the engine room, stokehold, and bunkers were flooded. Around 10 A.M. under the personal supervision of the captain, preparations were begun for getting out the port life boats and the Chief Officer was ordered to attend to the preparing and getting out of the starboard boats. At this time the *Vestris* had a list of approximately thirty degrees with her starboard deck under water and port rail high above the sea, and even with the davits swung out to their limits, the boats on the port side would drag along the ship's side while being lowered into the water, and it was a long and difficult operation to get them into the water without tipping them and spilling out the women and children who were placed in these boats, which were the first to be gotten over the ship's side, or without damaging the boats. The boats on the starboard side were gotten out and lowered into the water safely with the exception of one, No. 9, which was up-ended and swamped.

Many of the crew escaped into the starboard boats by jumping out into the sea and swimming to them or by swarming over the davits and down the falls into the boats. Boats Nos. 4 and 6 were never released from the falls and went down with the ship. No. 8 was damaged during the lowering and repaired, and although she got clear of the ship she was swamped shortly after she was lowered into the water. These were all [port side] boats and their passengers were women and children. At about 2:30 P. M. the *Vestris* capsized and sank in Lat. 37° 38' N, Long. 70° 23', which is about two hundred miles off the Virginia Capes. The captain was the last one to leave the ship, passing down her side into the sea just before she sank, and was lost. The steamships *American Shipper, Myriam, Berlin,* and the USS *Wyoming* arrived at the place of the sinking of the *Vestris* during Monday night and Tuesday, and rescued two hundred and thirteen persons from the life boats and wreckage, bringing some of the survivors to New York; others to Norfolk, Virginia.

II. The petitioners' contention is that they are entitled to exoneration under the Harter Act, 46 Mason's USC, secs. 190-95, as to the cargo on the ground that they had exercised due diligence to make the *Vestris* seaworthy on sailing and that the loss was due to perils of the sea or faults in navigation or in the management of the vessel. Petitioners also contend that if liability exists they are entitled, under the Limitation of Liability Act, 46 Mason's USC, sec. 183, to have it limited to its interest in the vessel and pending freight as it was occasioned without its privity or knowledge. The burden of proving the exercise of due diligence is upon the petitioners who seek the benefits of sec. 3 of the Harter Act, *Wildcroft*, 201 US 378, and in proceedings for limiting liability the burden of establishing its lack of privity rests on the petitioners. *In re Reichert Towing Line*, 251 Fed. 214, cert. denied 248 US 565. The following faults are charged by the claimants against the petitioners:

III. 1. Overloading; 2. Open and defective hatches in the shelter deck and sides; 3. Failure to determine the stability of the *Vestris* after the installation of the refrigerating machinery and insulation; 4. Failure to instruct the master and the chief engineer of the danger in pumping out the double bottom tanks; 5. Negligence in waiting until 9:56 A. M. Monday, November 12th, to send out the S.O.S. call; 6. Negligence in respect to care of passengers after the S.O.S. signal was sent out.

IV. In the effort to find the cause of the capsizing of the *Vestris* it is necessary to go back to the time when the first permanent list appeared, determine the cause of that and examine the sequence of events and conditions that developed from then on until the end. The history of the event has been narrated by a great number of witnesses. On many points they do not agree, and it is quite natural that the observation and the recollection of some of them would be affected by the mental and physical strain which they underwent. The testimony of others was tinged with a personal interest or that of their employers in the outcome of the litigation. But while there is doubt as to some of the details, the material facts have been established with a reasonable degree of certainty.

V. I think the weight of the testimony regarding the weather conditions justifies the conclusion that the weather alone, which was not exceptionally severe for that season of the year, is not an adequate explanation of the loss of the *Vestris* and that her loss was the result of a combination of conditions and events, no one of which alone would have caused it. Testimony too voluminous to analyze here in detail has been offered regarding the weather and the experiences of other vessels in the general vicinity of the *Vestris*. There is considerable variation in the descriptions of the weather and sea. All agree, however, that the wind increased during Sunday afternoon and was at its height Sunday evening and subsided soon after midnight; that on Monday morning the wind had gone down and the weather was fair, although there was a heavy swell. Third Officer Welland, who was called as a witness by the petitioners, testified that at 7.30 P. M. Sunday the wind was between force 9 and 10 on the Beaufort scale and that during the evening it increased to about force 10, with a high sea running. Welland's testimony impressed me as being reliable and after considering all the testimony in this respect, I think his statement as to weather conditions is substantially correct. Officers from some of the other vessels in the general neighborhood of the *Vestris*, although none were very near, testified that the wind reached force 12 Beaufort scale. Whether they are right in this or not, no other vessel—and some of them were small—suffered any serious damage. US Weather Bureau Officials who regularly prepare weather charts from radio reports from vessels and other sources described it as a severe Atlantic storm but not a hurricane. The weight of the testimony justifies the conclusion that the storm which the *Vestris* encountered was no more severe than is reasonably to be anticipated at that season of the year on a voyage from New York to South America; it was not extraordinary and should not have caused serious difficulty for a stable well-found ship. It does not fully account for the loss of the *Vestris* which in my judgment was due to her not being in proper condition to pass through it safely as which she otherwise would have.

VI. The testimony regarding the draft of the *Vestris* when she sailed from New York is inconsistent and because of its importance, I endeavored to exercise particular care in weighing the character of the witnesses on this point and the quality of their testimony. They were all, at least at the time of the event, in the service of the petitioners either on the *Vestris* or upon shore. I also realize that it is practically impossible even for experienced men to observe with accuracy the draft of a vessel without using a box-like device which shelters the marks on the vessel from the movement of the water, for the numerals on the vessel were marked with lines six inches apart and the distance of the water was estimated usually from the point above it. But while these approximate drafts were not accurate, it is probable that the estimates were not unfavorable to the vessel, as the owners were apparently content to accept them. And on a number of previous voyages the drafts reported to the managers of the line had exceeded the assigned draft, and on one voyage, to the extent of 14¼ inches.

The log of the *Vestris* went down with her, but Welland, the Third Officer, testified the draft entered in the log was aft 27 feet 11 inches. This was the draft reported by Watson, the Second Officer. The draft of 27 feet 11 inches was the draft from which the pilots "chit" was made and upon which the pilot charges were based, although this charge would have been the same whether it was 27 feet 11 inches or 27 feet 9 inches. Third Officer Welland could not remember the forward draft. Welland has left the services of the company to go into business and has no interest and his testimony impressed me as being reliable, and I accept the figure of 27 feet 11 inches as being correct for her draft aft. Varying statements by Captain Heasley, petitioner's Assistant Marine Superintendent in New York, regarding the draft leaves his testimony in this respect open to doubt. In order to arrive at the forward draft I have averaged the forward drafts as reported by Captain Heasley of 26 feet 8 inches, by Lloyd, petitioners' Harbor Master, of 26 feet 8 inches, and by Regan, petitioners' wharfinger, of 26 feet 7 inches, and accept the result of 26 feet 8 inches as her forward draft. Having a forward draft of 26 feet 8 inches and a draft of 27 feet 11 inches aft, her mean draft was 27 feet 9½ inches. This, however, was her draft at her pier in the Hudson River which is regarded as fresh water in figuring drafts, and a deduction of 4 inches from these figures is allowed for salt water. So that I find the mean average of the *Vestris* for this voyage was 26 feet 11½ inches. I certainly do not think it was less than this.

In 1912, upon leaving her builders in Belfast, load line or "Plimsoll" marks were affixed to the *Vestris* in accordance with the British Merchant Shipping Acts and the draft standards of the British Board of Trade. Under this standard she was assigned a salt water draft of 26 feet 3¼ inches for winter and 26 feet 9¾ inches for summer voyages. So that the *Vestris* sailed on this winter voyage loaded 8¼ inches below her assigned load line marks for winter voyages, and even below her summer marks by 1¾ inches. While in my judgment the British load line statute and regulations which required the affixing of the "Plimsoll" marks upon a vessel before leaving a British port and provided that a vessel loaded below those marks would be regarded as unsafe, is not applicable as a statute to the *Vestris* on a voyage from New York, an American port (for the reasons stated in my opinion filed September 15, 1931 A. M. C. 1553), such standards may be and should be taken into consideration together with all other facts and other accepted standards if there be any in determining whether she was overloaded or not. That vessels could only be loaded with safety to a certain depth has been known and recognized by sea-faring nations for many centuries. Load line marks upon vessels were established and officially recognized by the members of the Hanseatic League. Upon the dissolution of the League and the termination of its agreements for carrying on commerce between its members, recognition by one nation of another's load line marks ceased and the marking fell into disuse for a time. However, in 1875 Samuel Plimsoll, who was interested in the welfare of British seamen and owners of British cargoes, and realizing that many seamen, as well as cargoes, had been lost as a result of the overloading of vessels, introduced in the British Parliament in the year 1875 a bill for establishing and marking upon British vessels the point which indicates the limit to which a vessel may be safely loaded and giving to the British Board of Trade board powers for the proper carrying out of these purposes. In 1876 the bill became a law and was included in the British Merchant Marine Act. It was the purpose of the statute to have marks painted on the sides of the vessel indicating the point, ascertained by recognized standards, below which she could not be immersed without depriving her of a sufficient percentage of reserve buoyancy to insure her safety. In 1876 the plan of determining and indicating the load line mark, or "Plimsoll" mark as it is called, came into general use in England and was soon followed by several other sea-faring nations. The "British Shipping Act of 1894" required the "Plimsoll" or load line mark be determined and placed upon all British ships sailing from British ports, and under the British "Merchant Shipping Act of 1906" the rules were made to apply to foreign ships while in British ports "as they apply to British ships." The standard of the British Board of Trade relating to the determination and the affixing of the load line are the result of many years of experience and the history of the mariners and the ship builders of England is one to inspire confidence in their knowledge and skill in such matters. The weight of the testimony is that it represents a correct measure of safety. Standards for load-line marks substantially similar to those of the British are used by many of the seafaring nations, and there is a mutual recognition of each others' freeboard certificates. The American Bureau in assigning load-line marks follows standards which are essentially like those in force in Great Britain although it was not until September, 1930, that the United States enacted a statute relating to load-line marks.

Including the voyages made by the *Vestris* in the period from January, 1926 up to but not including the fatal voyage of November, 1928, it appears that she had made eight winter voyages when she loaded below her marks to the extent of 8¾, 8¼, 4¼, 1¼, 7¼, 14¾, ¾ and 6¼ inches respectively and had made seven summer voyages when she was loaded below her marks 4¼, 1¼, 16¼, 19¼, 2¼, 4¼ and 3¼ inches respectively. The ship's logs of each voyage or extracts from them which included the drafts were regularly received by the petitioners at their home office in Liverpool after each voyage, so that they were aware of what was going on. These voyages are referred to by the claimants as proof that she was overloaded on several voyages and that the owners knew that it was being done, and are referred to by the petitioners as indicating that it was safe for her to load below her marks as she was on this last voyage.

It is questionable whether the fact that the *Vestris* had made port on these several voyages although loaded below her Plimsoll mark indicated that it was a safe practice, or whether under favorable conditions she might reach her destination in spite of her being loaded too deeply. It does not appear what the weather or other conditions were on those voyages. But we do know that on this voyage when the weather was no more severe than might reasonably be expected at that season of the year, she met with disaster.

In 1926 the British Board of Trade, upon learning that *Vestris* had sailed from New York on her voyage on the 29th of May loaded below her assigned marks, wrote to the petitioners for an explanation and were advised by the petitioners that although the *Vestris* had left the pier loaded below her marks, she anchored at Liberty (in the New York Harbor) and pumped out enough water ballast to bring her up to her proper marks. This explanation was accepted by the British Board of Trade, but it now appears that this explanation was incorrect, as an abstract of the log of the *Vestris* then in the petitioners' home office would have disclosed.

The *Vestris* was owned by the petitioner, Liverpool, Brazil and River Plate Steam Navigation Company, Ltd., which was a mere holding company as all of its stock was owned and all its ships were operated by the petitioner, Lamport & Holt, Ltd. Mr. Wheeler was petitioners' Marine Superintendent in New York. and Captain Heasley was Assistant Marine Superintendent, and both of them knew when the *Vestris* sailed that she was loaded below her marks, and their knowledge was equivalent to the corporation's knowledge. *Benjamin Noble*, 244 Fed. 95, affirmed 249 US 334. Moreover, extracts from her logs showing her overdrafts from 1925 to 1928 were regularly sent to the owners and the practice of loading her below her marks was so frequent and so long continued that the owners were chargeable with knowledge of it. *Bessie J.*. 268 Fed. 66, affirmed 276 Fed. 778; certiorari denied 258 U. 5. 620; *In re P. Sanford Ross*, 204 Fed. 248; *In re Jeremiah Smith & Sons, Inc.*, 193 Fed. 395. I think it is clear that the petitioners have not sustained the burden of establishing the lack of privity and knowledge which rests upon them under the Limitation of Liability Acts. *In re Reich Towing Line, supra.*

I think the fair conclusion from all the testimony is that the methods of determining and assigning load line marks of the *Vestris* were in accordance with the generally accepted practice and that they indicated the point beyond which she could be loaded without incurring extra risk. In support of this conclusion there is also the testimony of a well-qualified expert.

It may be that loading the *Vestris* 8¼ inches beyond her marks was not the direct cause of her listing, but it did bring the openings through which water might enter nearer to the water and as several her openings were not water-tight, water came in. While a reduction of only 8¼ inches in freeboard might alone not have been particularly serious, this loss added to the reduction in her freeboard due to the fact that she was a tender vessel and listed between four to five degrees early Sunday morning with a consequent loss of freeboard on her starboard side of about 6 inches for each degree of list, resulted in bringing her openings on that side between

32¼ and 36¼ inches nearer the water at that time. The vital significance and the value of a few inches of freeboard is reflected in the action of Lloyd's and the British authorities by assigning to the *Vestris*, after careful consideration, 6½ inches less freeboard for summer voyages than voyages in the winter season. Loading below the marks also reduced her reserve buoyancy by several hundred tons. When a vessel bends over the maximum stability is arrived at about the time or shortly after the time that the edge of her deck begins to go under water, so that the less freeboard the sooner that point is reached, and consequently her range of stability becomes diminished.

VII. When she left New York on Saturday, the *Vestris* was on an even keel; between midnight and two o'clock on Sunday morning a slight list to starboard was noticed; between that time and five o'clock in the morning she had taken a permanent list to starboard of two to four degrees; at 7.30 A.M. the inclinometer indicated a list of between four to five degrees. There was merely a fresh northeasterly breeze during this period, and, although it had increased at 7:30 A.M., the wireless report, which it was customary for vessels to send in to the US Weather Bureau, the *Vestris* reported the wind then at a force which, according to the code issued to the *Vestris* by the Bureau, was about 35 miles an hour. While there is conflicting testimony as to the severity of the weather during Sunday afternoon and evening, there is no doubt at all that up to Sunday forenoon the weather was moderately good and the only reasonable conclusion that can be drawn from the fact that the *Vestris* listed four or five degrees under the conditions existing before 7:30 Sunday morning, in the absence of other cause, is that she was a tender ship and lacked stability. This conclusion is confirmed by the testimony of Welland, her Third Officer, and others. Welland testified on cross-examination as follows:

"Q. As a matter of fact you knew this was a tender ship, didn't you? A. Yes.

"Q. She had had lists before when you had been on her, hadn't she? A. She had experienced lists."

Referring now to the fact that the *Vestris* was a tender ship, which I think is fully established, it may well be that her tenderness alone would not have occasioned her loss and that she would have made port, but being tender she was more susceptible to other troubles which might become serious in the extreme; if there was a continued incursion of water and it was not gotten rid of, it would run to the lower or starboard side and increase her list to such an extent that some of the openings on that side would be submerged, and if she was lacking in adequate freeboard there would be greater danger of this occurring. Looking back over the situation as it progressed, it appears that due to her reduced freeboard caused by her being too deeply loaded and also by her having a list, defective openings in her sides and deck were, during the storm, brought nearer to the sea; also that water came in through them and much of it was not gotten rid of and ran to her lower side, and this gradually increased the list until her starboard half doors were submerged—these would begin to submerge if she listed 11 degrees—and later other openings, which finally resulted in her capsizing. While the explanation that a sanitary pipe may have broken, allowing water to be discharged into the ship and caused the list, is advanced—the state of the weather or the sea was not such during the period as to break a sound pipe. It is also quite apparent that if a ship has a list and she ships water through any source and it is not gotten rid of, it will flow to the lower side and tend to increase the list. And I think that there is no doubt that the fundamental cause of the increasing list and foundering of the Vestris was due to sea water which came aboard her through various openings and remained in the vessel, although the list was aggravated by pumping the water out of some of the tanks on the starboard side.

It is therefore necessary to search for the sources by which the water entered and to ascertain whether the petitioners were at fault with respect to them or any of them. And it is of course true that the deeper she was loaded the less freeboard she had, and the less freeboard—the nearer her openings would be to the sea. Openings on the lower side would be brought still nearer to the surface of the water by the listing of the vessel. If a vessel has a list it is evident that the incursion of water will increase the list because it will tend to congregate on the side of the vessel which is lower, and this results in reducing the freeboard on that side and bringing the openings on that side nearer to sea, so that she is exposed to further danger if water gets through any of them. The proof is that there were several openings where water did enter which, under the circumstances, resulted in further increasing the list so that her half doors on the starboard side were gradually submerged, allowing water to enter around the bottom and the sides of the doors, causing her to list more and more until other openings were exposed, and finally causing her to capsize and sink.

While it is obviously impossible to measure the amount of water that entered the ship at the various openings, I think there is sufficient evidence upon which to base a reasonable estimate of the amount of water that came in and the effect of it. The principal places where water entered were the starboard booby hatch in the upper deck, the half doors in the sides, and the ash ejector.

VIII. The water which came through the booby hatch was particularly serious and perhaps was one of the principal contributing causes of the loss of the *Vestris*. Water which came in through the booby hatch ran down into the firemen's passage and the cross alleyway to the starboard shelter 'tween deck bunker and thence through hatches into the lower bunkers. If water ran into the coal it could not be reached by the pumps, or to be more accurate, the pumps could only get at water that filtered through the coal and was not absorbed by it. The testimony of Professor Bugbee, Associate Professor of Mining Engineering and Metallurgy at Massachusetts Institute of Technology, who conducted an experiment with coal similar to that in these bunkers, is that the coal would hold and absorb, or as he described it "entrain" water weighing about 16% of the weight of the coal if there was a free opportunity for draining, and as the drainage was prevented in some of these bunkers the coal might entrain as much as 31% in water. Water entering the booby hatch would have access to some 2300 tons of coal, so that 360 tons of water might be held in the bunkers before it reached the bilges. All of this water would be on one side, and as and if it entered, the list would increase and the margin of stability diminish; and as much of the added weight would be high up it would tend to emphasize its adverse effect, for it raised the center of gravity and reduced the ship's metacentric height. Her freeboard would of course be decreased and the openings on her sides, the half doors, etc. brought nearer to the water.

The *Vestris* was a well-deck vessel and the seas were breaking over it and flooding it, as they would be likely to do in rough weather. If the opening was not tightly closed and the water in the well was higher than the 18 inch sill of the booby hatch, as it must have been to carry away the door, it would enter and run down into the firemen's alleyway. The testimony as to when the water began to enter through the booby hatch and the amount which came in there is contradictory but my conclusion from it all is that during the period from Sunday morning on, water was coming in through this opening at frequent intervals and that a considerable amount of water accumulated in the upper starboard bunkers from this source, even before the door of the hatch was completely carried away, which occurred Sunday evening, and when that occurred a larger volume of water came in. There is testimony of several witnesses that there was water in the firemen's alleyway Sunday morning, and that late Sunday the water in this alleyway was splashing about and overflowing into the 'tween deck coal bunkers. There were two scuppers in this alleyway but they did not carry away the water which came there either because they were small or stopped up. Whether the water which began coming in through the booby hatch Sunday morning all came through spaces between the bottom of the door and the sill and around the sides; whether some of it came in when the door was opened by the crew in going back and forth, or whether the door was damaged and began to leak before it was entirely carried away, is not definitely shown, but there is a strong inference at least that it was a result of a combination of these conditions.

The two large hatches inside the shelter deck bunkers leading into the lower bunkers were not closed or battened down but were left open so that the coal might run by gravity to the bunkers below as the coal was being consumed. This allowed water to lodge in these bunkers where it could not be reached by the pumps. The starboard hatch in the cross alleyway was battened down Sunday forenoon and a small hatch in the corner of the shelter 'tween deck bunker was closed Sunday evening. Why the others were not battened down then is undoubtedly explained by the fact that the hatches and coverings were not available in this emergency. As originally constructed the hatches had hatch coamings and hatch covers which, when in place, prevented this very thing. It had been their practice since the time the vessel was received from the builders to leave these hatches open and both Mr. Wheeler, petitioners' Marine Superintendent in New York, and Captain Heasley, Assistant Marine Superintendent, were aware of it. There were covers for these hatches, but they were never used and were left on the deck of the bunkers and the coal was dumped in on top of them, so that the covers were buried under the coal and not available unless the coal was removed.

It is urged by claimants that means should have been provided for stopping the water at the upper deck either by having an effective door for the booby hatch opening or by the closing of the hatches in the bunkers of the upper deck. This was a necessary requirement. The petitioners contend that there was a due compliance with this necessity, but in view of what happened I cannot agree with them. In "Practical Shipbuilding" by Campbell Holms, Third Edition, page 71, which was referred to by both petitioners and claimants upon the trial as a recognized standard book on Naval Architecture—

"Class I closing appliances (iii superstructures) must be of steel or iron; when closed they must be as strong as the intact bulkhead and practically watertight. Neither they nor the means of securing them may be detachable from their positions, and the arrangement must he such that the openings may be closed from both sides. . . .

"Class 2 closing appliances may consist of hard-wood doors, at least 2 inches thick, not more than 30 inches wide having sills at least 15 inches above the deck. Or they may consist of cross shifting boards, fitted the full height of the opening in channels riveted to the bulkhead, having a thickness of 1 inch for each 15 inches span, with a coaming plate at least 15 inches high. . . . The deck within a superstructure closed by Class 2 appliances must be watertight. Hatchways and other openings in the deck must have coamings at least 9 inches high, and means of closing as effective as those required for hatchways on the after part of a superstructure deck (Art. 455). Doorways in machinery casings protected by such superstructure must have coamings at least 15 inches high.

"If there are no means of closing openings in the end bulkheads of a superstructure, all openings in the deck within the superstructure must be protected by coamings at least 18 inches high; and as the deck is here practically a weather deck, freezing ports must be provided for in the sides, having an area of at least 2 square feet for each 10 feet in the length of the superstructure—except in the case of short open forecastles."

The regulations referred to in Article 455 mentioned above require the same methods of closing by strong-backs, hatch covers and tarpaulins fastened with iron bands and wedges as on the open decks of cargo ships, except that the coaming in the shelter deck need be only 9 inches high instead of 18 or 24 inches as required for hatches of an open deck.

The petitioners admit that the *Vestris* should have had class 1 closing appliances on her superstructure but contend that the requirement for class 1 closing appliances applies only to "openings in the end bulkheads of superstructures." However, it seems to me that the opening of the booby hatch was in fact equivalent to an opening through the end of the bulkhead of the superstructure, and that the same reasons for having class 1 steel doors for closings on the superstructure apply with equal force to the opening of the booby hatch; therefore that either the class 1 closing appliance of steel should have been provided for the booby hatch or the hatches in the upper deck bunkers should have been closed. Either of these arrangements would have prevented water from entering and going down into the vital parts of the vessel. That these wooden doors were in an exposed place and liable to damage is further indicated by the fact that doors of the booby hatch similar to these had been carried away on a previous voyage.

The petitioners take the position that the owners had done all that they were required to do as they had "provided" covers for the hatches and were not responsible if those on the ship failed to use them. But when the *Vestris* sailed, her shelter 'tween deck bunkers were filled with several hundred tons of coal and it was practically impossible in a sudden emergency to get at the hatches buried beneath the coal and batten them down. Cases holding a vessel unseaworthy where she sails with open ports amid cargo stowed in such a manner that they cannot be gotten at afford analogy. See *International Nav. Co.* vs. *Fair & Bailey Mfg. Co.*, 181 US 218; Manitoba, 104 Fed. 145. In the *International Navigation* case, Mr. Chief Justice FULLER said at p. 225:

"We do not think that a ship owner exercises due diligence within the meaning of the Act by merely furnishing proper structure and equipment, for the diligence required is diligence to make the ship in all respects seaworthy, and that, in our judgment, means due diligence on the part of all the owners' servants in the use of the equipment before the commencement of the voyage and until it is actually commenced."

The practice of leaving these hatches uncovered had continued for many years and both the Marine Superintendent and the Assistant Marine Superintendent in New York knew it, so that the petitioners are chargeable with privity and knowledge of the fact.

IX. "From the beginning of the day" Sunday, water was coming in at times through the opening around the starboard half doors where the packing of spun yarn and white lead had been washed away, and although the ship's carpenter put in new packing several times, the seas kept washing it away. He said that "That door stays off a little from the ship's side, a little like that—the door (indicating) . . . and when the sea is working in there it washes all the spun yarn out." And his following testimony is significant not only in respect to the condition of this half door, but in connection with the reduced freeboard:

"Q. Don't you think that at the same time the water was washing out the packing in the coal ports? A. No.

"Q. Why not? A. The coal ports have been tighter than the half door. The coal ports—they were not so big as the half door and in between the ship's side and the half door is more room for water.

"Q. What do you mean by more room for the water? A. Well, the sea goes more on the doors and in that way the coal port is a little bit higher than the half door.

"Q. Not very much higher? A. Not much.

"Q. How much higher? A. Well, about two feet."

From the plan of the *Vestris*, it appears that the opening for the half doors is 6 feet high and 4½ feet wide. The door is divided into two sections of half doors—an upper and lower. The coal ports are 3 feet wide and 2½ feet high, and the top of the coal port is on a line 2 feet above the top of the half door opening; the line of the bottom of the coal port is 6 inches below the top of the half door opening. The total amount of water which came in through this opening was under the conditions extremely serious in view of other conditions, although a greater quantity came here after the doors began to be submerged, which would have been when her list reached 11 degrees which occurred early Sunday evening. During the forenoon there was, at times, some two feet of water in the starboard end of the cross alleyway, part of which ran off into the bunkers and lower part of the vessel.

X. The starboard ash ejector began leaking very early Saturday morning and the water ran into the stokehold. At noon time, when it was stopped, it had filled the starboard bilges up to the floor of the engine room, which was about 3 feet above the tank tops. Part of this water was pumped out, but 18 inches remained in the afternoon and this was free surface water which would tend to run to her lower side and reduce her stability.

XI. At 6 o'clock Sunday morning it was discovered that water was coming in through an opening between the port half doors. The ship's carpenter who repaired it around 11 o'clock that morning testified that he could see about ½ inch of daylight between the upper and lower doors where they failed to meet. While some water entered at this point for several hours and added to the general situation, I do not think that the amount which could have come in through this opening could have been very much and the condition of these doors is not a ground for serious criticism.

XII. Although the main engine pumps were in operation from 9 o'clock Sunday morning on and the ash ejector pump was put in operation Sunday night, the water in the vessel increased, for more came in than these pumps could get rid of. The combined capacity of these pumps was limited to 180 tons an hour, as they all discharged through one 3½-inch pipe, and on two occasions during Sunday afternoon the ballast pump was used to pump out ballast tanks. The bilge injection pump was ordered into operation at 8 P. M. on Sunday. The use of this pump included the closing of the sea injection valve which supplies water from the sea to the circulating pump and thence through the condenser, also opening the bilge injection valve, so that water is taken from the bilges instead of the sea, and pumping it overboard through the condenser. However, when the bilge injection valve was opened and the sea injection valve was closed, the pump raced and did not work, and the use of it was abandoned.

XIII. The initial stability of the *Vestris*, either in light condition or as loaded, was unknown at the time she sailed. There are a number of factors which must be definitely and precisely established from which the metacentric height and initial stability of a vessel laden with cargo is computed, and it is now impossible to determine the metacentric height or her stability when she sailed, as her metacentric height light is unknown and the weight and exact distribution of the cargo can only be estimated and approximated. These estimates of the deadweight capacity of cargo, coal, water, stores, etc. vary and the total of these variances is so material

that the results are not reliable. Her metacentric height in light condition at
the time of the commencement of this voyage was negative as compared
with neutral at the time she left her builders' hands, as a result of the
refrigeration machinery and insulation subsequently installed, according to
computations made by Professor Jack which appear to be correct, but these
computations also are partly based upon the estimated weight and
metacentric height of this refrigeration machinery and insulation. Stability
is such a vital thing in a ship it would seem that the owners of the *Vestris*
were remiss, under the circumstances, in not definitely determining her
metacentric height, for she must have adequate metacentric height to
insure her stability and safety, but admittedly they did not know what it
was. When her builders conducted the usual test at the time she was built,
her metacentric height in light condition was neutral. But between 1912 and
1928 several hundred tons of additional refrigerating machinery and
insulation was installed. Correspondence passed between Lamport & Holt,
Ltd. in Liverpool and Workman, Clark & Co., builders of the *Vestris*, after
the refrigeration installation had been completed in New York in 1914,
which resulted in some 1400 tons being added to her weight, in which
reference was made to the fact that this would impair to some extent her
stability, as it undoubtedly would do. However, her metacentric height was
never re-determined either by attempted computation or by a new inclining
test, and it is now impossible to determine accurately what it was. Professor
Jack states that it would be little more than a guess to attempt to compute it
now; that in his judgment a much more nearly accurate result may be
obtained by a comparison with the *Vauban*, a sister ship, whose metacentric height had been determined by an inclining test in February, 1929, and
make allowances for the differences between the two vessels, than to
attempt it by independent computations with many factors unknown, and
that even by this method only the approximate metacentric height of the
Vestris in light condition may be ascertained.

The *Vestris* and the *Vauban* were alike in form, but the *Vestris* was built in
the Isherwood system, in which the strengthening beams ran fore and aft,
while with the *Vauban* the rib system was used in which the main
strengthening members ran transversely, with the result that the *Vestris* was
lighter by 322 tons than the *Vauban*. Professor Jack, after estimating the
height above the keel of the center of gravity (K. G.) of this 322 tons and

using the result of the inclining test of the *Vauban* in 1929 as a basis and making fair allowances for some differences in the insulation, etc. calculates that the metacentric height of the *Vestris* in light condition was .65 of a foot negative; that of the *Vauban* being .71 of a foot negative, this figure being kept full and hard pressed.

Both the *Vestris* and the *Vauban* were intended for carrying cargoes of chilled meats, and when they left the builders' hands some refrigerating machinery and insulation had been installed in each, although this work had progressed considerably further in the *Vauban* with the result that several hundred tons more had been placed aboard her than on the *Vestris*, and in 1914, when the refrigerating machinery and insulation was completed on the *Vestris*, some 1400 tons had been added to her since she left the yards of Workman, Clark & Co., her builders in Liverpool. The completed refrigerating machinery and the insulation on the two vessels were the same, except that the installation of the insulation on the *Vestris* required changes in details so as to conform it to her longitudinal frame, as the *Vauban* had a transverse frame. And it seems evident that the range of error is very much less in estimating the allowances which should be made for the small differences in the insulation of the two vessels due to the fact that one had cross beams and the other longitudinal beams, than in attempting to estimate the effect which this added 1400 tons would have on the original metacentric height of the *Vestris*.

Mr. Camps has endeavored to compute the metacentric height of the *Vestris* loaded as she sailed, independently of the *Vauban* test, and calculates that it was 1.77 of a foot positive. But it is evident that even the best effort of a highly qualified expert to now determine by mere computation what the metacentric height of the *Vestris* was in a loaded condition when she sailed on this voyage, is bound to be speculative, for not only was her metacentric height in light condition unknown, but also the weight and distribution of her cargo, water, coal, etc. can only be estimated, and the range of error to which these estimates are subject to would materially affect the result of the mathematical calculations.

I am quite convinced from the testimony in this case and from several authorities on the subject of shipbuilding, which I have studied with care, that the most satisfactory and the usual method of determining the metacentric height of a vessel of this type is by an inclining test, if that is possible, and if not, to use the figures obtained from an inclining test of a similar vessel and to calculate their differences. And it is significant that in 1921, when it decided that the stability of the *Vauban* should be determined, an actual inclining test was used. It seems to me that the nearest approach that can be made to determining the metacentric height of the *Vestris*, as she sailed loaded on this voyage, is to ascertain her light condition by comparison with the result obtained in the last inclining test of the *Vauban*, make allowances for the differences between the two vessels, and then, estimating the weights and distribution of the cargo, water, coal, etc., placed on board her, calculate what her metacentric height would have been under those conditions. Applying this method, it appears that the *Vestris* as loaded for this voyage would have had a metacentric height of 1.22 feet positive, assuming that her bottom tanks were full and hard pressed up. Most of the estimates upon which this calculation is based, with the exception of the light weight of the *Vestris*, were made up by the petitioners with the exception of the amount of coal, which was probably about 3,000 tons instead of 3,019 tons, and are subject to a material range of error not only as to the amount but also as to distribution of the weight. In making this calculation, a draft of 26 feet 11 inches was assumed, while I find that her draft was 26 feet 11½ inches. My own conclusion, after going into the matters in considerable detail, is that the *Vestris* in light condition (but with bottom tanks full and hard pressed up) had a metacentric height of approximately .65 feet negative; that the cargo that she carried tended to and did change her metacentric height from negative to positive; but that the exact amount of her positive metacentric height it is now impossible to determine owing to the lack of definite information as to the weights and distribution of cargo, coal, water, etc.; probably slightly under one foot positive, which is regarded by naval engineers and architects as sufficient provided that her deck openings and sides are watertight, although likely to make her tender; and that her small margin of metacentric height and other conditions resulted in her being a tender ship. This conclusion is confirmed by her actual performance early Sunday morning when she listed between two and five degrees and at that time the wind was not of sufficient force to affect a stable vessel.

XIV. The Chief Engineer, on orders from Captain Carey, had the starboard tanks Numbers 4, 5 and 2 pumped out. The pumping of No. 5 tank began shortly before 2 o'clock on Sunday afternoon. All the water which the pump could reach was gotten out by 5 o'clock; at 6 o'clock pumping began on No. 4 and was completed between 7 and 8 o'clock. Neither of these pumps had wing suctions, so all the water could not be gotten out of these tanks. At 4 o'clock Monday morning pumping was begun on No. 2 tank and finished around 7 o'clock. This tank was pumped dry as it had wing suctions. Instead of correcting the list to starboard, the pumping out of these tanks had the opposite effect. These tanks are located in the very bottom of the ship and the removal of the weight from that point raises the center of gravity of the ship and decreases the metacentric height (G. M.). Furthermore, as some of the tanks could not be completely pumped out, loose water was left in the tanks which would flow to the low side and aggravate the list. The petitioners concede that the pumping out of these tanks was wrong, but contend that this was merely a fault in seamanship and management by the vessel's master and officers, for which the owners are not responsible. However, it appears that when the refrigerating machinery was installed in the *Vestris* as previously mentioned, the builders of the *Vestris*, Workman, Clark & Co., in a written communication to the owners, stated:

"In the case of the *Vauban* and *Vestris* we find it necessary for trim and stability that Nos. 1, 2, 3, 4 and 6 double bottom tanks be filled, all the tanks with the exception of No. 6 fresh water tank being hard pressed. In addition to this, forepeak is filled, giving a total of 1035 tons of water ballast." **

** "We may mention that each of the vessels will be stable under any of the conditions we have stated, but it is understood that once the vessels get outside the river at Buenos Aires, the double bottom tanks will be filled to give sufficient stability for the voyage."

Despite this warning, the owners did not at any time inform Captain Carey, the master of the *Vestris*, or Mr. Adams, the Chief Engineer, as to the necessity of keeping filled these tanks of the vessel hard pressed up. The capacity of the tanks is: No. 1—60 tons; No. 2—280 tons; No. 3—300 tons; No. 4—220 tons; No. 5—270 tons, and No. 6—90 tons. Mr. Alfred Woods, the General Manager of the petitioners, admitted upon the trial that this information should have been conveyed to the master and officers of the *Vestris*. Failure by the owners to warn the officers of the *Vestris* of this requisite condition for the maintenance of her seaworthiness indicated, I think, a lack of due diligence on the part of the owners which their duty required. The pumping of these tanks was a contributing cause of the disaster as it tended to increase the list, and finally resulted in the foundering of the ship. In the *Elkton.*, 1931 A. M. C. 1040, 49 F. (2d) 700, at p. 701, Judge LEARNED HAND of the Circuit Court of Appeals of this Circuit, said:

"But existing unfitness, though it arise only from ill-handling of adequate equipment, must be so known at the outset; else she is unseaworthy, not being herself suited for her service, and not carrying a crew properly advised as to her needs. . . . the shipper . . . bears no loss until the owner has done his best to remove all risks except those inevitable upon the seas."

In *Standard Oil Co. vs. Clan Line*, L. R. [1924] Appl. Cas. 100 (J. L.) the master of the *Clan Gordon* while at sea ordered two of her water-ballast tanks emptied for the purpose of trimming her more by the stern. When they were nearly empty she heeled, turned over and became a total loss. Some years previous, after the loss of a sister ship by capsizing, the builders had warned the owners that when she was loaded as she was on this voyage, the water ballast tanks must be kept filled. This information was never given to her officers. It was held that the vessel was inherently unseaworthy in certain not improbable conditions unless special precautions were taken, and that the failure to warn the master constituted a breach of the ship owner's duty to exercise due diligence. LORD ATKINSON said:

"I think that the respondents, by leaving the captain of the *Clan Gordon* in ignorance of these instructions, by failing to bring them to his notice so that he would grasp and understand them, failed to discharge the duty they owed to the shippers of the cargo the vessel carried, and failed to use due diligence to make their ship seaworthy," page 123. *Schwan*, L. R. [1909] App. Cas. 430 (H. L.).

XV. The *Vestris* carried fourteen life boats, including one motor boat, with a total capacity of 800 persons, which were ample accommodations for the 325 persons on board. The life boats were of wood clinker built construction with copper air tanks that would keep them afloat although the boats filled with water, and with the exception of two which were extra boats, were under davits. These were "Martin" davits with "Mills" releasing gear which are generally regarded as satisfactory. Each boat carried the equipment of sails, oars, lanterns, bailer, food, water, etc., in conformity with the United States Inspection laws, although some equipment was lost in launching the boats and in other ways for which the owners are not responsible. The life boats had all been examined by the United States Inspectors a few days before she sailed, each boat having been lifted clear of the chocks with more than the official capacity number of persons in each boat. Lloyd's surveyor had also made a recent external examination of them and reported them in "very good condition." It is true that the boats were subjected to a tropical climate much of the time which has a tendency to dry them out and cause leaks, but it was the custom to keep water in them to prevent this. Several of the survivors who spent Monday night in the life boats have testified that there was water in them. This, I find, was chiefly the result of the waves coming over the boat's sides and the rain during the night, although water undoubtedly did come in through some of the seams, but no more than could have been gotten rid of by a reasonable amount of bailing. No. 8 boat had a hole stove in her bow while being launched. With the extreme list which the vessel then had, it was a lengthy and difficult job to launch a boat, so that the damage to No. 8 boat is readily explained. An inefficient effort was made to patch the hole with a piece of tin, but it did not prevent water from coming in, filling and capsizing the boat several times before being picked up. During this time many of her occupants were lost, as pathetically described by survivors. From what we now know, it seems to have been a mistake in judgment on the part of the captain who was in charge to have made use of this boat instead of trying to launch another to take its place. However, I find, after full consideration of all the evidence, that the *Vestris* carried a sufficient number of seaworthy and properly equipped life boats.

The *Vestris* followed the usual custom of allocating boat stations to members of the crew. Notices were posted in the cabins informing the passengers of their respective boats in the event of the ship meeting with disaster. On this occasion the system was not adopted. The captain himself took charge of the port boats and the chief officer was ordered to take charge of the starboard boats. The signal of one long blast, which was the call to boat stations, was never given because of the desire to avoid a panic. All passengers amid crew were, however, notified and apparently there was no general panic or exceptional confusion. As a result of the captain's effort to obey the mariners' law of chivalry, "Women and Children First," they were placed in the first boats which were attempted to be launched, with the lamentable result that, as two of the boats were not gotten clear of the vessel's sides and into the water before she sank, the women and children in them went down with the ship. This accounts for the unusually large proportion of women and children who were lost.

Since the disaster there has been much discussion and criticism of the fact that the port boats instead of those on the starboard side were used for passengers. This was an operation that was difficult and not free from danger, for the ship at this time had a list to starboard of upwards of 35 degrees and the port boats were 50 to 60 feet above the sea. Oars were used as skids in an attempt to hold the boats off from the side of the ship and extreme care was necessary to prevent a boat from overturning as it slid along the side of the ship, for the arm of the davit was, of course, not long enough to carry the lifeboat beyond the sloping side of the ship, lying in the position in which she then was. The starboard rail was under water and so that it would seem to have been better to have used her starboard boats for more of her passengers, and this seems to be demonstrated by the fact that all of the starboard boats were successfully launched with the exception of No. 9 which was swamped, and No. 13 which floated off; while only No. 8 and No. 10 were successfully launched on the port side. However, it was difficult and dangerous to get the passengers into the starboard boats either before or after they were launched. Passengers might have been put in the starboard boats while the boats were still in the chocks on the boat deck and swung out with the boats, but this was not free from danger. Members of the crew who escaped in the starboard boats either climbed out on the davits and dropped into the boats after they were launched or jumped in to the sea and swam to the boats. To launch the port boats with the passengers over the port side or on the starboard side was a difficult and serious problem—both hazardous, and although it may be that the master decided unwisely, it is interesting to note that upon the torpedoing of the *Lusitania* in 1915 and the attempt of the officers in charge to launch some of her boats with passengers from the high port side, some criticism of it was made later in the litigation which followed, and the court held that under the conditions at the time of the sinking which were not dissimilar from those with which the officers on the *Vestris* were confronted, those in charge could not be regarded as negligent even though it proved to have been extremely hazardous.

Captain Carey, the master of the *Vestris*, who was faced with the actual situation, was a man of many years' experience at sea, a goodly number of which he had been master on passenger ships, and the facts do not justify a conclusion that he acted negligently in deciding or in his attempt to carry out his plan of launching the port lifeboats. On the contrary, from the time he took personal charge of the work of loading them and getting them overboard until he went down with his ship, he was devoting all his efforts to saving the passengers before she sank, and he met this last crisis with the courage and the fidelity traditional with men of the sea.

XVI. The *Vestris* was equipped with radio apparatus and three wireless operators provided continuous watch. On Monday morning, between 8 and 8:30 E.S.T. o'clock, pursuant to an order from Captain Carey, the *Vestris*'s position was obtained by wireless and plotted and at 8:37 the alarm signal "C. Q." was dispatched. At 8:47 the Tuckerton Station rebroadcast the message as follows:

"All ships in the vicinity of *Vestris* please stand by and keep close watch for any distress messages may develop trouble."

["Stand by" as used here, meant to be listening in for the next message.]

At 9:56 an "S.O.S" message was sent out and was followed by the words "Requires immediate assistance please dangerous list going over all time starboard."

At 11:04 the *Vestris* repeated the "S.O.S." call. The two "S.O.S." messages were received by fifty-eight ships; six of which were then within a radius of 100 miles of the *Vestris*; thirteen within a radius of 150 miles, and twenty-six within a radius of 200 miles.

From an examination of a chart of these vessels it is apparent that if the "S.O.S." message had been sent between five and six hours sooner, one or more of them probably could have reached the *Vestris* before she went down, and rescued her passengers.

After the lurch at 7:30 on Sunday evening, the *Vestris* had a list of ten to fifteen degrees, and at midnight it had increased to fifteen degrees; at four in the morning, in spite of everything that had been done, the list was about twenty degrees and was increasing. There were two feet of water on the starboard side of the engine room. The fires of the starboard boiler had been put out by the water and they were having trouble with the pumps. It must have been apparent sometime after midnight Sunday and before four o'clock Monday morning, that the *Vestris* was in a desperate condition and that ships in the vicinity should be summoned to which her passengers could be transferred. Aside from the high degree of care which the captain of the ship was required to exercise as a common carrier of passengers, ordinary prudence demanded that every effort should have been made by that time to obtain assistance. Why Captain Carey neglected to summon aid sooner will probably never be known. Certainly he should have done so and it can only be assumed that he failed to fully realize the great danger to his passengers and ship.

Claimants cite the following paragraph contained in a book of "Instructions to Masters " issued by the steamship line to its masters as an authorization or encouragement for its masters to unduly risk the safety of its passengers and ships.

"In the case of a serious disaster happening to one of the vessels of this line while at sea, the master must in the first instance carefully consider the actual amount of the peril there may be for the lives of those under his charge and then judge whether he will be justified or not in fighting his own way unaided to the nearest port. His being able to succeed in this will always be considered a matter of high recommendation to him as a master." P. 48

However, such a conclusion seems a doubtful one.

The agents of the line got their first information that the *Vestris* was in distress when they heard that she had sent the general "S.O.S." message, and wirelessed Captain Carey as follows:

"November 12, 1928 10:28

Captain 'Vestris'

Wire immediately your trouble

LAMPORT."

To which the captain replied:

"November 12, 1928. 11 AM

Hove to from noon yesterday. During night developed 32 degree list to starboard. Deck under water. Ship lying on beam ends. Impossible to proceed anywhere. Sea moderately rough.

CAREY."

Upon receipt of this information, Lamport & Holt, the agents of the line in New York, sent the following wireless message to the *Vestris* at 11:27 A.M.: "United States Destroyer *Davis* proceeding to your assistance." The representatives of the owners also chartered a large ocean-going salvage tug and sent her to the aid of the *Vestris*. But although both vessels proceeded with all reasonable expedition, they arrived at the scene of the disaster too late; the *Vestris* had disappeared and the survivors had been rescued by other vessels.

XVII. While it is true that certificates of condition had been obtained for the *Vestris* which, in the absence of anything to the contrary are entitled to more or less weight, they must be valued in the light of what the actual conditions proved to be. For instance, certificates do not alter the fact that the port half doors were in such condition that on Sunday one could see half an inch of daylight between the upper and lower doors; and the starboard half doors were sprung and could not be closed tight; and that nothing had happened since the beginning of the voyage which would have affected good doors. The certificates should also be read in the light of other conditions which are referred to. Moreover, Lloyd's regulations required that the hatches in the shelter 'tween bunkers should be capable of being made watertight, and Lloyd's surveyors as well as the United States Inspectors were unaware that these would be left open on sailing. It does not appear that the United States Inspection Service considered or tested the stability of the *Vestris*; its inspection was principally concerned with engines, equipment, life boats, life preservers and general conditions. And it is important to keep in mind that the Lloyd's certificates were issued upon the express condition that the *Vestris* could not be loaded below her "Plimsoll" marks and that the others were undoubtedly issued upon the assumption that she would not.

XVIII. Two distinguished naval architects and marine engineers testified respecting the seaworthiness of the *Vestris* upon her departure upon this voyage, Mr. Harold Edward Joscelyn Camps of London, England, called by the petitioners, and Mr. James Robertson Jack of Scotland, now Professor of Naval Architecture and Marine Engineering at the Massachusetts Institute of Technology, called by the claimants. These experts, after being informed of all the known facts, disagree in respect to her being seaworthy; Mr. Camps testified that in his opinion she was seaworthy, and Professor Jack that she was not, as she was so deeply loaded that her openings were brought so dangerously near to the water that she should have had a steel closing for the booby hatch or the hatches in the upper deck should have been made watertight, and that she lacked a safe margin of stability and reserve buoyancy. With due deference for the views of both of these experts, my own conclusions, after very careful consideration, is that the facts seem to me to support the opinion of Professor Jack, not alone because of what actually did occur, but because it seems to me apparent, from the history of the disaster, that a principal contributing cause of it was the overloading which allowed incursion of water through defective openings and which increasingly affected her list and finally resulted in her foundering; also that her margin of stability and reserve buoyancy was not sufficient to carry her through a situation which was reasonably to be anticipated.

XIX. In an effort to account for the disaster the petitioners advanced the theory at the trial that it was due to a broken pipe, or pipes, leading through the bunker space; possibly the four-inch salt water overflow pipe running through the shelter 'tween deck bunker space from the sanitary tank to the boat deck which discharged through the ship's side a short distance above the upper deck; or possibly one of the three pipes leading through the hanging bunker below the upper deck discharging through the ship's side, each of which had a storm valve directly connected to the skin of the ship. Two of these three pipes leading through the hanging bunker discharge through the ship's side below the half door in the cross alleyway; one was a 3½ inch scupper to the cross alleyway itself; the second was a 5 inch soil pipe from the "W. C." [water closet: bathroom] on the shelter deck; the third was a 1½ inch scupper from the firemen's passage. This third pipe was in the immediate neighborhood of the coal chute, [therefore] being about midway between the soil pipe discharge and the firemen's alleyway scupper discharge. There were also 19 other pipes discharging through the shelter 'tween deck bunker space. Previous to sailing, these pipes had been examined and found in good condition. The theory that one of these pipes must have broken and permitted loose water to escape into the ship is largely based upon the petitioners' assumption that at 7:30 Monday morning the *Vestris* had a list of 25 degrees and that the estimated water from known sources would account for a list of only 21 degrees. However, if a reasonable margin of error be allowed in the estimates of the testimony as to the degree of the list at that time and the estimates as to the amount of water that entered through disclosed sources, the theory does not rest upon a very firm foundation. Particularly as the water entering the ship by the known sources could, it is conceded, have produced a list of 21 degrees and with this list the leaking half doors would almost entirely be submerged and would have brought the forward edge of the shelter deck almost to the waterline. This explanation, coupled with the testimony that the pipes were in good condition before the *Vestris* sailed, does not account for the gradual list which began around Saturday night when the weather was not sufficiently violent to cause good pipes to break, and all these openings were well above the waterline. This theory also fails to explain the large amount of water in the firemen's passages, cross alleyway, amid shelter deck 'tween bunker amid other parts of the upper deck. It also appears unlikely that one of these larger pipes could have broken and discharged this large amount of water without some of the officers at least suspecting that this had occurred and investigating, but no one did. The conclusion that the incursion of the water was not due to one of these broken pipes is confirmed by the knowledge that both Second Engineer Forsythe and Third Engineer Blue had been on the *Vestris* on a previous trip when one of the pipes now referred to—a 5 inch toilet discharge pipe in the after cross bunker, did break and they immediately recognized where the water was coming from, and exactly what had happened, and promptly took means to get at the pipe and repair it.

XX. I find that the sinking of the *Vestris* was due to a combination of causes including the fact that she was a tender [unbalanced or with insufficient ballast] vessel and over loaded, which reduced her buoyancy and stability, and which brought her defective openings nearer to the water, allowing water to enter; the failure of the owners to notify her master of the necessity of which they were aware of keeping her bottom tanks filled and which were pumped out by him; and that the owners have not sustained the burden of showing the exercise of due diligence to make her seaworthy in these respects, nor have they established lack of privity or knowledge of these conditions, which certainly contributed to the loss. The owners therefore cannot avoid the consequences. *Edwin I. Morrison,* 153 US 199; *Clark* vs. *Barnwell,* 53 US 272, at p. 280.

For the reasons above set forth, I find that the petitioners are liable for the losses resulting from the foundering of the *Vestris* and that their petition for limitation of liability must be denied, with the usual reference to a commissioner as to damages. If the foregoing be deemed insufficient compliance with Admiralty Rule 46½, Findings and Conclusions may be settled on notice.[1]

Obscure nautical terms (several found in this appendix)

Bilged: Bottom of boat damaged so as to allow water to come in.
Booby hatch: A sliding hatch or cover.
Clinker-built: a boat built with overlapping hull planks.
Davit: A crane, often working in pairs and usually made of steel, used to lower things like lifeboats over the side of a ship.
Falls: Rigging used to lower a lifeboat. A fall is a single line holding one end of a lifeboat. Since they are always used in pairs, they are called "falls."
Fishplate: a steel plate used to lap a butt joint.
Kentledge: weights used as permanent high-density ballast.
Metacentric height: The distance between the center of gravity of a ship and its metacenter (a quantity difficult to determine). A large metacentric height makes a ship more stable, but lessens the roll period, which makes it uncomfortable for passengers; tradeoffs are needed.
On beam ends: A ship listing at an angle of 45° or more.
Rhumb line: A path derived from a defined initial bearing.
Tender ship: A ship that is intrinsically unstable—top heavy.
Warfinger: The owner or keeper of a wharf.
Wearing ship: To turn a ship's stern through the wind so as to reverse the direction of the wind with respect to the ship.

Appendix B:
Vestris Analysis by K.C. Barnaby

This analysis of the *Vestris* sinking was written by Kenneth Cloves Barnaby, OBE, Honorary Vice-President of the Royal Institution of Naval Architects. These excerpts are taken from his book *Some Ship Disasters and Their Causes*, originally published in 1968 (Hutchinson) with the first American edition published in 1970 (A.S. Barnes and Company). Barnaby brings out some interesting facts about the *Vestris* disaster that are not common knowledge.

We begin with Barnaby's discussion of the situation aboard the *Vestris* on Sunday, November 11, 1928:

> Soon after noon, as the ship was steering wildly owing to the increasing sea and wind on the quarter, Captain Carey allowed the ship to lie in the trough of the sea with the engines stopped, except that the starboard engine was put ahead from time to time and the rudder to port with the object of bringing the ship somewhat up to the wind. Carey, in his message to his owners, stated that he was "lying hove-to." This, however, is not the proper meaning of hove-to, which is to lie meeting the wind with engines stopped, or nearly stopped. Carey did not survive to explain his actions, but several of them seem most unwise. Unless a ship is extremely high powered, the propellers have very little effect in turning the ship's head as they have such a small leverage. Similarly a rudder has no effect at all on a stopped ship, unless it is in the direct race of a screw. Carey was thus employing quite inadequate means when endeavoring to bring the ship "up to the wind." He had any amount of sea room and one naturally wonders why he did not adopt sailing ship tactics and "wear" the ship round, i.e., bring her stern to the wind. This

would have had several advantages. The water-soaked coal in the starboard bunkers would be much heavier than that in the port bunkers and would have remained so for a long time. This would have offset to a considerable extent loose water which would have crossed to the port side as soon as the wind came on the starboard side of the ship. Carey seemed to have been so confident that he could right the ship by pumping out starboard ballast tanks that he neglected a simple maneuver that would most probably have saved his ship. With an extremely tender ship that has little or no metacentric height when upright, it is of course useless to list her on the other side as she merely flops over to a similar angle. Carey presumably did not think of the waterlogged coal, or not in sufficient time. It would have been too late when the water in the starboard stokehold had reached the starboard boiler.

This is truly a startling revelation. Note that water did not reach the starboard boiler until about four o'clock Monday morning, so Captain Carey had a window of opportunity of some fifteen hours and forty minutes to effect the maneuvers described by Barnaby. This was certainly more than sufficient time to have done these things had the captain been so inclined. Yet the idea never seems to have entered his mind.

Barnaby continues his discussion with some of the events of the next day, Monday, November 12, the day the *Vestris* sank.

At 8:37 A.M. on Monday the alarm signal CQ was sent out, and at 9:56 A.M. the SOS signal. The position sent out was substantially correct. . . .

This statement by Barnaby is true. The rescue ships made the error of assuming that the debris field from the wreck would stay in the same place for many hours. This error contributed greatly to the loss of life, since rescue craft were at the SOS position as early as eight o'clock Monday night but did not find any lifeboats or survivors until about four o'clock the next morning. Meanwhile, an unknown number of persons died in the turbulent seas as they were carried away by the Gulf Stream.

Barnaby continues his discussion of Monday's events:

> From 10 A.M. until the ship foundered at about 2:30 P.M., Carey and his officers were working at the boats. Carey's order to start with the port boats, the weather boats, made things much more difficult owing to the severe list. It seems a quite unnecessary order as the *Vestris* carried fourteen lifeboats with a total capacity of eight hundred persons and had less than four hundred people on board at the time. The third-class passengers had been taken to the boat deck, the first and second-class had either been sent or had found their way to the port side of the promenade deck. The reason for starting with the port boats was no doubt the ease with which they could be loaded from the promenade deck. With a listed ship, the boats swing far outboard but they could have been loaded from the boat deck. The women and children were all put in the port boats, [Numbers] 4, 6, and 8. This traditional custom was and is a grave mistake. Now that ships carry "boats for all," families should be kept together. Apparently Carey ordered all the women and children to be placed in the weather boats as a precaution, and he fully expected rescue ships to arrive before the ship foundered. This did not happen, and although the port boats [Numbers] 4, 6, 8, and 10 were lowered to within ten to fifteen feet from the water, only [Numbers] 8 and 10 were actually put in the water. [Number] 8 was damaged during lowering and according to the official account was swamped soon after getting water-borne. [Number] 4 was never released from the falls and went down with the ship. The falls of [Number] 6 were cut, but she was stove in and capsized when the ship sank. [Number 2 remained on the chocks and went down with the ship.]
>
> . . .

The one satisfactory feature of the disaster was the complete order and discipline that prevailed at all times. It was only at 1:45 P.M., that is about forty-five minutes before the ship foundered, that, owing to the conditions in the stokehold and engine room, the engineers came on deck. Chief Engineer Adams was the last to leave. A large number of the Barbadian crew were saved by means of the starboard boats. Many of these men jumped from the ship and were then picked up by the boats. Some swarmed over the davits and down the falls. . . .

. . .

The foundering of the *Vestris* had come as a shock to the general public and to the seafaring community. It seemed almost incredible that a large liner of modern type, classed 100 A1 at Lloyd's and carrying a USA Passenger Certificate, should sink in a short-lived storm that had not worried other vessels and when only two days out from New York. The unpleasant and stark fact that, while 78 percent of the crew had been saved, only 47 percent of the passengers had been rescued, also required close examination.

Mr. Barnaby then considers the findings of the British Board of Trade inquiry into the disaster:

[The court] also thought that a general instruction to the masters of Lamport & Holt ships should be withdrawn. This read, "In the case of a serious disaster happening to one of the vessels of this Line, whilst at sea, the master must in the first instance carefully consider the actual amount of peril there may be for the lives of those under his charge and then judge whether he will be justified or not, in fighting his own way unaided to the nearest port. His being able to succeed in this will always be considered a matter of high recommendation to him as a master." The court very properly thought that "the captain's decision to avail himself of the SOS signal should not be affected by any such consideration."

Captain Carey certainly delayed much too long before sending out the SOS. At 4:00 A.M. on Monday the position was already hopeless. The ship was then heavily listed, water in large quantities was continuing to get into her, and according to the evidence the captain and the chief engineer could not determine the main source of the inflow. Had the SOS been sent out at that time, more than one ship would almost certainly have arrived at the scene before the *Vestris* sank. Vessels coming to her assistance would have had daylight in which to find the ship or boats, instead of having to search for them in the dark.

Instead of sending out a distress signal, Carey continued his quite unjustified optimism for nearly six hours. He ordered the starboard side of No. 2 ballast tank to be pumped out, blithely remarking, "That should bring her up." This was an imprudent order, because owing to the heavy list, the initial effect [was] to increase the list. Carey was acquitted of imprudence by the court on the ground that he had not the knowledge to realize this. He had, however, seen that pumping out No. 4 starboard [ballast tank] the previous evening had completely failed to "bring her up." The court's observations about Carey's handling of the boats were also extremely mild.

To this we should add: Any middle school student of general science would know enough to realize that pumping the water out of an underwater ballast tank and replacing it with air would destabilize a ship more than it already was. What nonsense that a ship's master would not have the knowledge of such a simple fact of physics!

Barnaby continues with Court of Inquiry findings:

The court did not attempt to explain Carey's quite unpardonable conduct in leaving the [Numbers] 4 and 6 boats hanging in the falls for so long laden with women and children. They should at least have been got clear of the vessel when the last engineers had come on deck and nothing further could be done to save the ship. The loss of the *Vestris* was a sad blow to the reputation of the British Mercantile Marine, and it was soon followed by the withdrawal of the two sister ships *Vandyck* and *Voltaire* from the American service and their conversion into cruising liners. It also seems to have been a quite unnecessary loss. As a small set-off against the loss of life, a number of contributions to increased safety at sea have resulted from the recommendations of the Court of Inquiry. These include better pumping arrangements, the enforcing of the load line rules in foreign ports, the supply of stability information to the ship's officers, and improved water-tightness of hatches and coal trunks.

Although the court listed a number of causes for the inflow of water into the *Vestris*, it is difficult to believe that these were really sufficient to account for the heavy list in the early stages of the disaster. A very reasonable theory was put forward by Mr. Little, M.I.N.A., and supported by both Sir John Biles and Mr. Edward Wilding. This assumed that the first serious inflow came through a broken sanitary discharge pipe of five-inch diameter. The court rejected this theory on several grounds, notably a greater draught than that assumed and the fact that the pumps were not working at the assumed efficiency. The author considers this rejection as far too cocksure, and he certainly prefers an opinion of Sir John Biles on a technical matter to that of the legal luminary, Mr. Butler Cole Aspinall.

It is of course most desirable that all sanitary and other discharges should run to the shell through spaces that are at all times fully accessible and not through coal bunkers or cargo spaces. . . .

So saying, Mr. Barnaby closes his section on the *Vestris* disaster with some remarks about cases where this precaution was not observed, with much mischief following. But these remarks are not germane to the wreck of the *Vestris*.

Appendix C

Letters Written to Judson G. Jackson

AFTER THE STEAMSHIP *Vestris* went down with the loss of over 100 lives, Judson G. Jackson asked for reports from known survivors of the wreck if they had any information about his parents, Dr. Ernest Jackson and Jannette Jackson, or his brother, Cary Jackson. All three were lost when the ship went to the bottom of the Atlantic on November 12, 1928.

Jackson received replies from six of the survivors: William Phipps Adams; Marion C. Batten, wife of race car driver Norman K. Batten, who was lost; Anne DeVore, wife of race car driver Earl F. DeVore, who was also lost; Campbell Kellman of Barbados; Dr. Ernst F. Lehner of Basel, Switzerland; and Captain Frederik Sorensen, who was traveling as a passenger on the *Vestris*.

All of these were two-page letters except the one from William Adams, which had five pages. Images of the original letters from these six persons are reproduced (in grayscale) on the next fourteen pages, in alphabetical order by surname. Also included are two pictures from the scrapbook of Judson Jackson, one of the sinking *Vestris* and the other of the two race car drivers, Batten and DeVore, who were lost after getting on lifeboats that sank.

WILLIAM P. ADAMS

Odebolt, Iowa Nov 30 19 28

Mr Judson Jackson
    Knoxville Tenn

Dear Sir :-

    Replying to your circular letter to the survivors of the Vestris :-

    Sunday morning about ten I was reading a book on the "port" or high side of the ship, in the ladies saloon when the vessel gave a hard lurch & every thing & every body on that side, including the chairs tables rugs etc slid down to the low side of the room — at that time a very hard storm & high sea was running — It seems that Mr Jackson had been sitting near me on the high side also & after we crawled out of the jamb of furniture &c Mr Each got a big overstuffed arm chair & put it on the high side of the Vestris — Knowing we couldnt slide any further, and entered into a general conversation — I told him of my expected fishing trip to some islands off the Chili coast from Valpairaiso, and he told me of his life & work in the Baptist missions in Brazil — I said to myself "Thats a nice man" — & havent changed my mind. After about an hour of this I went to my room and do not remember seeing him again — nor on the

First page of a letter from William Adams to Judson Jackson[1]

promenade deck the next morning – nor in
any of the Small boats in the afternoon but I
was at the after End of the deck which had a
list about like this / so I could do nothing but
hold to the rail or stauction until I got into a
boat with my friend (he is 72 + I am 66) + did not
walk about the deck nor again go below for that reason

It may well have been that he thought his
Chances were best on the lee side of the ship – but
a raft with 50 people on it + 2 boats full that
were launched last on that side were caught
by the superstructure of the Ship when it finally
rolled over — + I presume many if not all were lost
— The next boat to ours on the port side was
very leaky + filled + sank shortly after being launched

I do not think that Dr Jackson was in the
Same boat with his wife as this was not allowed +
the only order I heard given + the only officer I saw
that day was the Capt ordering a man out of the
boat that Contained the women — he was saved + his
wife (on their honeymoon) was lost —— That boat was
the first loaded but, for some reason was the last
to leave the ship — in fact, it never did leave
the hull of the ship for, when we were about

Second page of Adams letter to Jackson

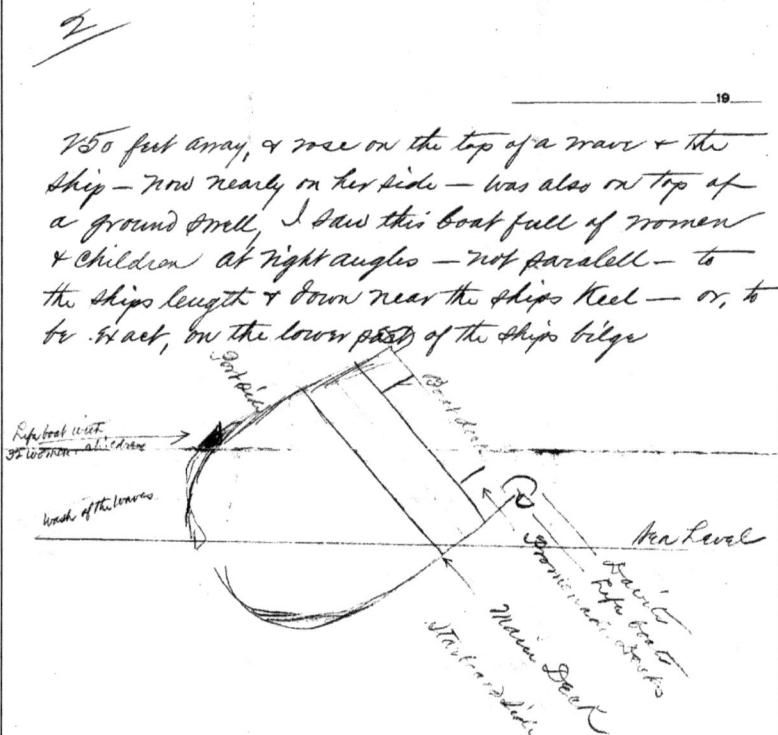

William P. Adams

2

19

150 feet away, & rose on the top of a wave & the
ship — now nearly on her side — was also on top of
a ground swell, I saw this boat full of women
& children at right angles — not paralell — to
the ships length & down near the ships keel — or, to
be exact, on the lower part of the ships bilge

Life boat with
35 women & children

Wash of the waves

Sea Level

& Several persons were standing on the side of the ship trying to push
this boat into the water — Then our boat sank into the hollow of a wave &
the next time I got on top — & the ship also rose on a swell — that boat was
gone & the ship had rolled onto her side — & filled & sank in 2 or 3 minutes
Last summer I broke my arm & macerated my shoulder & when I got
on the American shipper I saw a sailor there with his arm in a sling

Third page of Adams letter to Jackson

and, in fellow sympathy, I got to talking with him & asked what
had happened to his arm — It was in a cast & sling — He said he
was in charge of the boat with the women in it I have spoken of
& that as they were trying to work the boat down into the water a
cargo boom or a davit or some thing very heavy fell down on
the boat & crushed it &, in passing struck his shoulder (his
forearm was amputated late,) — The occupants, more or less
injured, I assume, were thrown in the water along side the hull
of the ship & many must have drowned at that time & no doubt
the rest were sucked down when the ship sank a few minutes after
— The boat next to ours was very leaky & filled & sank
very soon after it was launched full of men

The sea was rough — 12 to 18 ft waves & a good ground
swell running & you will realize that only at certain
times could one see much in such a sea way — & some
of these things I saw & some I heard of from others when
I got on the American Shipper the next day — It was
about 2 o'clock PM when we got away from the ship &
as the boats were rocking about & in danger of collision, we
made all haste to get away from them & also from the
suck of the sea as the ship sank, as well as floating debris that would be there soon after — We had 47 in our
boat & could take no more — 6 or 7 were white — I did
most of the "talking" but it didnt do much good — The men
were laborers — not sailors and, any how, it was every one for himself
— It took 4 to bail our boat all the time —— How I did
wish for my Colt Automatic (in my trunk) & a couple of feet of
garden hose to make those negroes bail when I said bail

Fourth page of Adams letter to Jackson

WILLIAM P. ADAMS

③

_____ 19__

& "Sit down" when some one saw a "brilliancy" that night on the horizon & called out "Ship"!! & nearly tipped the boat over — We had no officer in our boat — nor did I see any on the ship but they may have been getting out boats on the starboard side

I have a winter home at Miami Beach Fla & know Mr John W Green — a lawyer — of Knoxville & we intend to drive to Fla in 10 days or 2 weeks & shall call on Mr & Mrs Green on our way down &, at that time I will ~~try to call on you & shall then be able to tell you~~ more as my stenographer is ill & I have never been a good penman — but, I do not know more of Mr Jackson than I have written & Im sorry to write that.

Yours very truly
W P Adams,

Fifth page of Adams letter to Jackson

262 Central Park West,
New York city, N.Y.
Dec. 27, 1928.

Dear Jackson Family,

Forgive me for not
writing sooner but have been so wretched
at times that writing has been out of
the question.

To answer your pathetic question,
can only tell you that the only time
I saw Dr and Mrs Jackson and Cary
was on Monday morning Nov. 12th when
everyone was gathered in the smoking
Room with their life belts on awaiting
what we knew not then. Dr. Jackson read
a few verses from his Bible which were
appreciated by every one. They were not
in my life boat, that I know of, and I
never saw them again.

This has been one great tragedy for
us all. I lost my dear husband after holding
on to him till the help was almost at hand.
Hoping you get more information,
    Very Sincerely   Marion C. Batten

Letter from Marion C. Batten to Judson Jackson[2]

(written on very heavy card stock)

1710 Jerome Ave
Brooklyn, N. Y.
12/ 12/ '28.

Judson Jackson,
Knoxville, Tennessee.

Dear Mr. Jackson;—
         Am very sorry
that I am unable
to give you any
information about
Dr. and Mrs. Jackson
who were on the
ill fated "Vestris."
         I saw them
at the dinner table

First page of a letter from Anne DeVore to Judson Jackson[3]

Sat. evening, but on Sunday I was in my state room all day as the sea was very rough — and there were only few who were able to be up and around. Then I saw them at noon on Monday just before we got into the life boats, but I did not see them getting into the boat, as there was such a rush the last few moments.

    I remember the boy very well, as I had admired him, he was such a nice looking boy, and so sweet to his mother.

    You have my sincere sympathy in your great loss, I can understand your sufferings, as I lost my dear husband in the same disaster.

    Very Sincerely,
        Mrs. Earl DeVore.

Second page of DeVore letter to Jackson

Dear Mr Jackson & Family,

I received your letter which was addressed to my brother in Upper Montclair & redirected to me in Barbados B.W.J.

First I must offer you all my deepest sympathy and I can only hope & pray they may be found.

I can remember Dr Jackson & family quite well. I was sitting in the Music Room the Dr had a Brazillian nut which he said was quite difficult to obtain and was good for the stomach. He gave my sister Mrs Shreve's & myself some which was quite bitter but beneficial. I afterward heard him speaking to another gentleman of his Church and work.

I am sorry to say I did not notice them when the ship

First page of a letter from Campbell Kellman to Judson Jackson[4]

went down
         I did not have anything to
say to Mrs Jackson but was struck
by her kind face. Gay I met on
the stairs several times on Sunday
evidently fetching something for
his father & mother.
         Again offering you
my sympathy I remain
                    Sincerely yrs.
                         C.Kellman.

Rock Hall
     St. Peter
    Barbados
         B. W. I.

Second page of Kellman letter to Jackson

Pointe a Pierre, 20th December 1928

Mr. Judson Jackson,

    Box 104, Knoxville, Tenessee, U.S.A.

Dear Mr. Jackson,

    Permit me first to express to you all my heartfelt sympathy. I recollect having seen your brother Cary during the whole of Sunday and to have spoken a few words to him. I do not remember having seen your father and mother that day; apparently they kept to their cabin, or to the social hall .

I can vividly recall the picture of your parents and Cary sitting in the Smoking Room on Monday morning. They were very quiet and composed. We even conversed on some trifling matters in the early morning hours. When it became certain that we would have to take to the boats, your mother sent Cary down to the cabin to fetch some of her things, which he brought up in a small hand-bag. After that both your father and Cary did splendid work in collecting life-belts from the cabins, which they distributed among these people, who were too frightened to get them themselves. When everybody in the smoke room had their life belts, your father started praying in a a low, but firm voice, conforting so the little group, which by now was augmented by some coloured women and children from the second and third class. I then left the smoke room, but I recollect having seen your mother embark in a life-boat, either No. 6 or No8,on the port side. I do not remember having seen your father. When I embarked into life boat No.10,

First page of a letter from Dr. Ernst Lehner to Judson Jackson[5]

there were about 4 or 5 passengers still on deck (port-side),
amongst them Cary. He was the last in a line which was slowly
moving along the railing towards the place where we had to get
on to the rope ladder. He never embarked in our boat, however.
Whether he realised himself that boat No.10 was already overlaa-
ded and tried to get across to the starbord boats, or whether he
was ordered to do so, nobody will ever know.

I have told you what little I know, because, in your great
grief, it may mean a little consolation to you to hear how
splendidly your parents and Cary behaved in the face of disaster,
and how they were bent on service to their fellow passengers to
the last moment. Personally I was particularly struck by the
~~heroism of Cary. I shall never forget him, and am proud to have~~
known him.

                    Yours sincerely

                              E.Lehner

Second page of Lehner letter to Jackson

Hotel St. George
CLARK STREET
BROOKLYN NEW YORK CITY

1200 ROOMS
800 BATHS

PHONE
MAIN 10000

Nov. 28 - 1928.

Dear Sir:

Sorry to say I did not get to know
Mr. & Mrs. Jackson while on the Vestris.
But when I look at their pictures I can
remember seeing them before we abandoned
the ship. Not so with the boy. ~~If I had~~
at least not that I can remember.

But Mr. and Mrs. Jacksons picture has remained
in my memory. I feel quite sure on
account of the following: Shortly before
abandoning the ship a number of passengers
were gathered in the smoking room.
Mr. Jackson was seated with his wife.
Suddenly Mr. Jackson spoke up in a

First page of a letter from Capt. Frederik Sorensen to Judson Jackson[6]

loud clear voice something like this:
" We must now pray and trust in God
to help us out.,
That was the last time I can remember
seeing them. Had it not been for that
sentence they would probably not have
have attracted any of my attention.
Regretting that this is all the information
I can give you and assuring you of my
deepest sympathy.
                    I am yours truly
                    J. Sorensen.

Second page of Sorensen letter to Jackson

This remarkable picture of the Vestris disaster shows the ill-fated ship battling through seas and listing sharply—a listing which began, according to passengers, even as the vessel went down New York Harbor at the beginning of its trip of death. The photograph was taken by a passenger on the Vestris a short time before the ship floundered off the Virginia Capes.

Now given up for missing are two celebrated automobile racers, Earl F. Devore, above, and Norman K. Batten, both of Los Angeles.
Mrs. Devore tells a tragic story of members of a Vestris' lifeboat crew refusing aid to her husband while he was struggling in the water.
Had letters from their wives.

From the Scrapbook of Judson G. Jackson[7]

The photo above shows the badly listing *Vestris* perhaps an hour or two before she turned turtle and sank. It may be one of the photos taken by Fred Hanson.

The photos on the left show the two race car drivers who died in the disaster. Earl DeVore is above, and Norman Batten is below. The note at the bottom of the newspaper article was written by Judson Jackson. In it he states that he received letters from the wives of the two race car drivers. Marion Batten and Anne DeVore were both among the survivors and lost their husbands in tragic circumstances.

Endnotes

Most of the information referred to in this book can be found in the transcripts of the Tuttle hearings or the Custom House inquiry headed by Dickerson N. Hoover, Inspector General of the Steamship Inspection Service. Two volumes of these transcripts are held by the New York Public Library. We believe the sources listed below are the most widely available versions of the testimony and accounts given by the survivors of the *Vestris* disaster.

Chapter One

1. *New York Times*, November 15, 1928, page 3.[†] "Tragedy Termed 'Murder.'" Interview at dockside.

2. Ibid.

3. — November 15, 1928, page 1. "'*Vestris* Will Sink,' Friends Told Quiros." Written account by William Carlos Quiros.

4. — November 15, 1928, page 3. "Tragedy Termed 'Murder'." Interview at dockside, direct quotation.

5. — November 22, 1928, page 14. "*Vestris* Officers Refused to Man Lifeboats, Chief Says; He Admits Chaos in Crisis." Testimony from the Custom House inquiry held by the Steamship Inspection Service. Source herein called "the Hoover inquiry."

6. — November 21, 1928, page 16. "Men in Crew Say *Vestris* Leaked on Four Voyages; Was Fit, Inspectors Insist." Testimony from the Tuttle hearings quoted at the Custom House inquiry held by the Steamship Inspection Service.

7. — November 25, 1928, page 28. "*Vestris* Inquiry Stirs British Shipping Men." British "expert" was not named in the article; the quotation was taken from "British newspapers."

[†] Newspaper page numbers given are where the quotation or text appears. The headline given may—and usually does—appear on an earlier page.

8. — November 16, 1928, page 2. "Six Survivors on Stand." Testimony at the Tuttle hearings.

9. Ibid.

10. *Time*, National Affairs, November 26, 1928, page number unknown. "Catastrophe: *Vestris.*"

11. *New York Times*, November 23, 1928, page 2. "*Vestris* Put Boats Off on Wrong Side, Ex-Captain Asserts." Testimony at Tuttle hearings. Paraphrased testimony, no direct quotes from article.

12. Ibid.

13. — November 16, 1928, page 3. "Two Victims Died in Arms of Wives." Interview on board the USS *Wyoming* at dockside.

Chapter Two

General note for Chapter Two: Most of the facts related in this chapter that are not directly sourced to endnotes can be found in the *New York Times* editions of November 15 through 19, 1928. They contain lengthy articles of testimony taken from the Tuttle hearings.

1. *New York Times*, November 28, 1928, page 4. "Carey Was Misled, Says *Vestris* Mate." Testimony at the Tuttle hearings.

2. — November 16, 1928, page 1. "Stoker Lays Wreck to Open Coal Door." Interview; location not stated in article.

3. *Journal of Commerce, The.* Report of the [British] Board of Trade Inquiry into the Loss of the *Vestris* and Findings of the Court, page 50. Testimony. Source herein called the "Board of Trade inquiry."

4. *New York Times*, November 18, 1928, page 1. "Four Leaks Sank *Vestris*, Engineer Says." Tuttle hearings; testimony paraphrased and quoted.

5. — November 16, 1928, page 1. "Stoker Lays Wreck to Open Coal Door." Interview; location not stated in article.

6. Board of Trade inquiry, page 77. Testimony.

7. *New York Times*, November 28, 1928, page 4. "Carey Was Misled, Says *Vestris* Mate." Testimony at the Tuttle hearings.

Chapter Three

1. Chicago *Southtown Economist*, November 20, 1928, page 1. "Doctor Tells of Terror as *Vestris* Sank." Interview by reporter John C. Metcalfe. Only quotes by Dr. Groman—which are facts—are included; no original material by Metcalfe is included herein.

2. *Time*, National Affairs, November 26, 1928, page number unknown. "Catastrophe: *Vestris*."

3. *Baltimore News*, November 15, 1928, page 1. "Crew of *Vestris* Took Best Boats, Says Witness." Testimony by Fred W. Puppe at the Tuttle hearings, distributed by the Associated Press to evening newspapers and datelined "New York, Nov. 15."

4. Letter from William Phipps Adams to Judson Jackson dated November 30, 1928. Courtesy of Jackson Estate. Complete text is in Appendix C.

5. Kalafus, Jim, December 18, 2006. "My God the boat is leaving us." *Gare Maritime* on *Encyclopedia Titanica*. At URL:
http://www. encyclopedia-titanica.org/my-god-the-boat-is-leaving-us-vestris.html

6. *New York Times*, November 15, 1928, page 1. "'*Vestris* Will Sink,' Friends Told Quiros." Written account by William Carlos Quiros.

7. *New York Times*, November 15, 1928, page 1. "Tells Vivid Story of Growing Alarm." Written account by Dr. Ernst Lehner.

8. — November 15, 1928, page 3. "Tragedy Termed 'Murder.'" Interview at dockside.

9. Ibid.

10. Ibid.

11. Ibid.

12. — November 15, 1928, page 4. "Survivors Tell of Heroes." Interview at dockside.

13. — November 15, 1928, page 1. "Tells Vivid Story of Growing Alarm." Written account by Dr. Ernst Lehner.

14. — November 15, 1928, page 1. "Drifted 22 Hours with Woman in Sea." Paul Dana, as told to Lorena Hickok, AP staff writer. Used with permission of The Associated Press Copyright © 2015. All rights reserved.

15. — November 15, 1928, page 1. "'*Vestris* Will Sink,' Friends Told Quiros." Written account by William Carlos Quiros.

16. — November 15, 1928, page 3. "Tragedy Termed 'Murder.'" Joint interview at dockside, in which both men contributed remarks that were not individually identified, with one exception.

17. Chicago *Southtown Economist*, November 20, 1928, page 1. "Doctor Tells of Terror as *Vestris* Sank." Interview by reporter John C. Metcalfe. All directly quoted material.

Chapter Four

1. Chicago *Southtown Economist*, November 20, 1928, page 1. "Doctor Tells of Terror as *Vestris* Sank." Interview by reporter John C. Metcalfe, direct quotation.

2. — November 20, 1928, pp. 1–2. "Doctor Tells of Terror as *Vestris* Sank." Interview by reporter John C. Metcalfe. Only direct quotes of Dr. Groman included here.

3. *Journal of Commerce, The*. Report of the Board of Trade Inquiry into the Loss of the *Vestris* and Findings of the Court, page 112.

4. *New York Times*, November 16, 1928, page 2. "Six Survivors on Stand." Testimony at the Tuttle hearings.

5. *Time*, National Affairs, November 26, 1928, page number unknown. "Catastrophe: *Vestris*."

6. Appendix A, pp. 198–199.

7. *New York Times*, November 20, 1928, page 2. "Captain of *Vestris*, in Daze at Wreck, Said 'Damn the Crew,' Stoker Declares." Interview.

Chapter Five

1. *Time*, National Affairs, November 26, 1928, page number unknown. "Catastrophe: *Vestris*."

2. Ibid.

3. Letter from Captain Frederik Sorensen to Judson Jackson dated November 27, 1928. Courtesy of Jackson Estate.

4. Letter from Dr. Ernst Lehner to Judson Jackson dated December 20, 1928. Courtesy of Jackson Estate.

5. *New York Times*, November 16, 1928, page 2. "Six Survivors on Stand." Testimony at the Tuttle hearings.

6. Ibid.

7. — November 17, 1928, page 2. "Vestris Men Fail to Explain SOS Delay; Tuttle Charges Witnesses with Evasion on Earlier Message 'We May Need Aid.'" Testimony at the Tuttle hearings.

8. Ibid.

9. *Journal of Commerce, The*. Board of Trade inquiry, pp. 86–87.

10. *New York Times*, November 23, 1928, page 2. "Ship's Ex-Captain on Stand." Testimony at the Tuttle hearings.

11. Parshall, Ardis E. "En Route to the Field: Missionaries Aboard the S.S. *Vestris*, 1928." Keepapitchinin—The Mormon History Blog: http://www.keepapitchinin.org/2009/04/03/en-route-to-the-field-missionaries-aboard-the-ss-vestris-1928/

12. Letter from William Phipps Adams to Judson Jackson dated November 30, 1928. Courtesy of Jackson Estate. In Appendix C.

13. Chicago *Southtown Economist*, November 20, 1928, page 2. "Doctor Tells of Terror as *Vestris* Sank." Interview by reporter John C. Metcalfe. Direct quotes from Dr. Groman.

14. *New York Times*, November 19, 1928, page 2. "*Vestris* Seamen Eager To Leave New York." Interview at Seaman's Church Institute in New York City.

15. — November 16, 1928, page 2. "Six Survivors on Stand." Testimony at the Tuttle hearings.

16. — November 15, 1928, pp. 1 & 15 (AP). "Drifted 22 Hours with Woman in Sea." Interview by AP reporter Lorena Hickok. Used with permission of The Associated Press Copyright © 2015. All rights reserved.

17. *Journal of Commerce, The.* Board of Trade inquiry, page 85. Testimony of Thomas Robinson, bedroom steward.

18. — Board of Trade inquiry, page 85. Testimony of Thomas Connor, head waiter in the first-class salon.

19. — Board of Trade inquiry, page 86. Testimony of James McCulloch, a laundryman on the *Vestris*.

20. — Board of Trade inquiry, page 88. Testimony of Warwick Roberts, boatswain on the *Vestris*.

21. *Time*, National Affairs, November 26, 1928, page number unknown. "Catastrophe: *Vestris*."

22. *New York Times*, November 15, 1928, page 5. "Survivors Tell Of Heroes." Interview.

23. *Journal of Commerce, The.* Board of Trade inquiry, page 51. Testimony of Leslie Watson, second officer on the *Vestris*.

24. *New York Times*, November 15, 1928, page 5. "Survivors Tell Of Heroes." Interview.

25. Ibid.

26. — November 15, 1928, page 1. "French Ship Brings 57 from the *Vestris*." Interview.

27. — November 15, 1928, page 5. "Tragedy Termed 'Murder.'" Interview.

28. — November 15, 1928, page 13. "Crew Lays Sinking to Coal Door Leak." Interview.

29. — November 14, 1928, page 1. "Survivor Tells Of The Disaster; Had To Jump From Crowded Lifeboat." Account by William Carlos Quiros written for *La Nacion* of Buenos Aires.

30. — November 20, 1928, page 2. "Captain Kept Peril Secret; *Vestris* said, 'Nothing New' 5 Hours Before the SOS." Testimony given by Arthur J. Costigan of RCA concerning the logs of radio messages sent and received by the *Vestris* on Monday, November 12, 1928, as recorded by the Tuckerton, New Jersey, shore station. Editorial comments are those of Mr. Costigan.

31. — November 20, 1928, page 2. "Captain Kept Peril Secret; *Vestris* said, 'Nothing New' 5 Hours Before the SOS." Testimony given by Arthur J. Costigan of RCA about radio bearings made on the *Vestris* on Monday, November 12, 1928, as recorded in the logs of the Chatham, Massachusetts, radio station.

32. *Journal of Commerce, The*. Board of Trade inquiry, page 86.

Chapter Six

1. *New York Times*, November 15, 1928, page 10. "Officers Describe Rescue On The Radio." "Two officers and a surgeon of the American Merchant liner *American Shipper*, which played a heroic part in the rescue of survivors of the sunken Lamport & Holt liner *Vestris*, described their part in the drama in addresses broadcast last night over Station WOR and the Columbia Broadcasting System, through the courtesy of the Olster Radio Corporation."

2. — November 16, 1928, page 2. "Six Survivors on Stand." Testimony at the Tuttle hearings.

3. *The Rockford* (IL) *Daily Register-Gazette,* November 14, 1928, page unknown. "Twenty-seven Women Missing on *Vestris*."

4. *New York Times*, November 16, 1928, page 2. "Six Survivors on Stand." Testimony at the Tuttle hearings.

5. — November 15, 1928, page 4. "Survivors Tell of Heroes." Dockside interview, direct quotation.

6. — November 21, 1928, page 16. "Swears Ship Leaked Before." Testimony by Thomas Connor at the Tuttle hearings.

7. — November 15, 1928, page 2. "Survivors on Three Ships." Dockside interview, direct quotation.

8. — November 15, 1928, page 2. "Survivors on Three Ships." Dockside interview of Anne DeVore (surname misspelled "Devore").

9. — November 15, 1928, page 5. "Tragedy Termed 'Murder.'" Interview, direct quotation.

10. Kalafus, Jim, December 18, 2006. "My God the boat is leaving us." *Gare Maritime* on *Encyclopedia Titanica*. At URL: http://www. encyclopedia-titanica.org/my-god-the-boat-is-leaving-us-vestris.html

11. *New York Times*, November 15, 1928, page 4. "Survivors Tell of Heroes." Dockside interview, direct quotation.

12. *Journal of Commerce, The*. Board of Trade inquiry, pp. 72–73. Testimony of Gustav Wohld.

13. *New York Times*, November 15, 1928, page 4. "Survivors Tell of Heroes." Dockside interview, direct quotation.

14. *Journal of Commerce, The*. Board of Trade inquiry, page 46. Testimony of Frank William Johnson, chief officer of the Vestris.

15. — Board of Trade inquiry, page 49. Testimony of Frank William Johnson, chief officer of the Vestris.

16. *New York Times*, November 15, 1928, page 3. "Tragedy Termed 'Murder.'" Dockside interview, direct quotation.

17. — November 22, 1928, page 14. "*Vestris* Officers Refused to Man Boats, Says Chief; He Admits Chaos in Crisis." Testimony at the Custom House hearing under Inspector General D.N. Hoover.

18. — November 15, 1928, pp. 4–5. "Survivors Tell Of Heroes." Interview aboard the *American Shipper*.

19. *Rockford* (IL) *Daily Register-Gazette*, November 14, 1928, page 2. "27 Women Missing on *Vestris*." Interview.

20. *New York Times*, November 15, 1928, page 5. "Survivors Tell of Heroes." Dockside interview.

21. — November 16, 1928, page 2. "Six Survivors on Stand." Testimony at Tuttle hearings. "Dead paint" is paint that is old and oxidized, rough, and probably cracked in many places.

22. *Journal of Commerce, The*. Board of Trade inquiry, pp. 88–89. Testimony of Warwick Roberts, boatswain on the *Vestris*.

23. *New York Times*, November 22, 1928, page 14. "*Vestris* Officers Refused to Man Boats, Says Chief; He Admits Chaos in Crisis." Sub-heading: "Disagree with Inspectors." Testimony at the Custom House hearing under Inspector General Dickerson N. Hoover.

24. National Oceanic and Atmospheric Administration, Earth Research Laboratory, Sunrise/Sunset Calculator; online at URL: http://www.esrl.noaa.gov/gmd/grad/solcalc/sunrise.html

25. *New York Times*, November 15, 1928, page 3. "Tragedy Termed 'Murder.'" Dockside interview, direct quotation.

26. — November 15, 1928, page 3. "Tragedy Termed 'Murder.'" Dockside interview, direct quotation. Named here as "Conrad F. Slaughter," his actual given name was Cline.

27. — November 15, 1928, page 4. "Passenger Says All Would Have Been Saved If SOS Had Been Sent Out At The Proper Time." Interview at Mr. Marvin's home in Montclair, New Jersey.

28. — November 22, 1928, page 14. "*Vestris* Officers Refused to Man Boats, Says Chief; He Admits Chaos in Crisis." Sub-heading: "Disagree with Inspectors." Testimony at the Custom House hearing under Inspector General Dickerson N. Hoover.

29. — November 22, 1928, page 14. "Vestris Officers Refused to Man Boats, Says Chief; He Admits Chaos in Crisis." Sub-heading:

"Disagree with Inspectors." Testimony by Edward Marvin at the Custom House hearing under Inspector General Dickerson N. Hoover.

30. — November 15, 1928, page 4. "Survivors Tell of Heroes." Dockside interview of Edward J. Walsh.

31. Chicago *Southtown Economist*, November 20, 1928, page 1. "Doctor Tells of Terror as *Vestris* Sank." Interview by reporter John C. Metcalfe.

32. Found during research in public records and now cannot be located. Neither author has any doubt as to the authenticity of this quote.

33. Letter from William Phipps Adams to Judson Jackson dated November 30, 1928. Courtesy of Jackson Estate. Complete text is in Appendix C.

34. New York Times, November 23, 1928, page 2. "Says Capt. Sorensen Assailed Ship's Crew." Interview.

35. — November 16, 1928, page 2. "Six Survivors on Stand." Testimony at the Tuttle hearings.

36. — November 15, 1928, page 2. "Man Who Saved 20 is a Reticent Hero." Paraphrased interview.

37. — November 15, 1928, page 3. "Tragedy Termed 'Murder.'" Interview at dockside.

38. — November 15, 1928, page 5. "Survivors Tell of Heroes." Bedside interview at St. Vincent's Hospital.

39. — November 15, 1928, page 3. "Tragedy Termed 'Murder.'" Interview at the Fifth Avenue Hotel.

40. Ibid.

41. *Kingston* (NY) *Daily Freeman*, November 14, 1928, page 1. "Captain Carey Criticized by Some Survivors." Dockside interview.

42. *New York Times*, November 15, 1928, page 2. "French Ship Brings 57 from the *Vestris*." Interview.

43. Parshall, Ardis E. "En Route to the Field: Missionaries Aboard the S.S. Vestris, 1928." Keepapitchinin—The Mormon History Blog:

http://www.keepapitchinin.org/2009/04/03/en-route-to-the-field-missionaries-aboard-the-ss-vestris-1928/

44. *New York Times*, November 15, 1928, page 5. "Survivors Tell Of Heroes." Interview of David Huish.

45. *Augusta* (GA) *Chronicle*, November 14, 1928, page 1. "206 Passengers of Steamer Vestris On Board Rescue Vessels Steaming Toward Home Ports." Associated Press report; direct quotes only.

46. *New York Times*, November 15, 1928, pp. 1 & 15. (AP) "Drifted 22 Hours with Woman in Sea." Interview by AP reporter Lorena Hickok. Used with permission of The Associated Press Copyright © 2015. All rights reserved.

47. *Rockford* (IL) *Morning Star*, November 18, 1928, page 26. "Identities Didn't Count When Pair Battled Sea." The article appeared under the dateline of "New York, Nov. 17." but no source is named. The PARS agency, which handles permissions for Associated Press material disclaims any copyright in this op-ed piece.

48. *New York Times*, November 15, 1928, page 4. "Survivors Tell of Heroes." Dockside interview with Clara Ball.

49. — November 15, 1928, pp. 1, 4. "Survivors Tell of Heroes." Dockside interview with Clara Ball.

50. — November 15, 1928, page 6. "Devore [*sic*] Went Down, Survivors Report." Quotation from an unnamed passenger on the *American Shipper*.

51. *Daily Mail, The* [London], December 6, 1928, page 11. "*Vestris* Disaster." Verbatim quote of remarks made by Capt. Forey of the *Myriam* on his arrival at Hull, England.

52. *Dallas Morning News*, November 16, 1928, pp. 1, 3. "Wife of Texan Tells of Rescue From *Vestris*; Mrs. Slaughter Says Captain Did His Best." Cf. *New York Times*, November 15, 1928, page 3. "Tells How Officer Rescued Her Twice."

53. *Rockford* [IL] *Daily Republic*, November 15, 1928, page 2. "Pretty Matron Tells Story of Escape From Sea."

54. *New York Times*, November 16, 1928, page 3. "Tells How Officer Rescued Her Twice." Interview.

55. — November 15, 1928, page 3. "Tragedy Termed 'Murder.'" Dockside interview.

56. *Journal of Commerce, The*. Board of Trade inquiry, page 51. Testimony.

57. *New York Times*, November 16, 1928, page 4. "Death Toll Now at 111." Interview conducted some time after survivors came ashore.

58. — November 21, 1928, page 19. "Swears Ship Leaked Before." Testimony at Tuttle hearings.

59. *Rockford* [IL] *Daily Register-Gazette*, November 14, 1928, page 2. "27 Women Missing on *Vestris*." Interview.

60. *New York Times*, November 15, 1928, pp. 1 & 6. "'*Vestris* Will Sink,' Friends Told Quiros." Written account by William Carlos Quiros, Chancellor of the Argentine Consulate in New York City.

61. — November 16, 1928, page 3. "Relief Provided for Rescued Crew." Dockside interview.

62. — November 15, 1928, page 3. "Tragedy Termed 'Murder.'" Dockside interview.

63. *Rockford* [IL] *Daily Republic*, November 14, 1928, pp. 1 & 2. "Score of Little Children Die as Boats Capsize." Quoted interview.

64. *New York Times*, November 18, 1928, page 5. "*Vestris* Survivor Seized For Theft." Paraphrased news article.

65. *Rockford* [IL] *Daily Republic*, November 14, 1928, page 2. "Score of Little Children Die as Boats Capsize." Paraphrased news article.

66. *Journal of Commerce, The*. Board of Trade inquiry, page 100. Testimony.

67. *Rockford* [IL] *Daily Republic*, November 15, 1928, page unknown (not in clipping file). "'Iron Man' of Sea Disaster Recounts Story." Direct interview with Carl Schmidt, location not specified.

68. *New York Times*, November 15, 1928, page 7. "Says Stokers Fought Over Rescue Ladders." Paraphrased news article.

69. — November 15, 1928, page 14. "Rescuing Skippers Are Modest Heroes." Dockside interview.

70. *Kingston* [NY] *Daily Freeman*, November 14, 1928, page 1. "Two Vessels Land With 148 Survivors of the *Vestris*." News article.

71. *New York Times*, November 16, 1928, page 3. "Two Victims Died in Arms of Wives." Interview on board the USS *Wyoming* at dockside.

72. — November 23, 1928, page 2. "Witnesses at *Vestris* Hearings Describe How Liner Filled." Testimony at Tuttle hearings.

73. Ibid.

74. *New York Times*, November 22, 1928, page 15. "Vestris Officers Refused to Man Boats, Says Chief; He Admits Chaos in Crisis." Testimony at the Custom House hearing under Inspector General Dickerson Naylor Hoover (brother of J. Edgar Hoover, who became in 1924 director of what would become, in 1935, the FBI).

75. Owosso, MI, *Argus-Press*, November 18, 1928; headline and page number unknown.

76. *New York Times*, November 18, 1928, page 3. "Wave Took Batten From Wife's Arms." Interview in Hotel Belmont.

77. Ibid.

78. Kalafus, Jim, December 18, 2006. "My God the boat is leaving us." *Gare Maritime* on *Encyclopedia Titanica*. At URL:
http://www.encyclopedia-titanica.org/my-god-the-boat-is-leaving-us-vestris.html

Chapter Seven

1. Cf. *Time*, National Affairs, November 26, 1928, page number unknown. "Catastrophe: *Vestris*."

2., 2a. *New York Times*, November 16, 1928, page 2. "Six Survivors on Stand." November 18, 1928, page 1. "Four Leaks Sank Vestris, Engineer says."

3. Cf. *Time*, "Catastrophe: *Vestris*."

4. *New York Times*, November 23, 1928, page 2. "*Vestris* Put Boats Off on Wrong Side, Ex-Captain Asserts." Paraphrased testimony at Tuttle hearings. Cf. *Time*, National Affairs, November 26, 1928, page number unknown. "Catastrophe: *Vestris*."

5. Ship's Nostalgia, "The *Vestris* Disaster," online at URL:
http://www.shipsnostalgia.com/guides/
Royal_Mail_Steam_Packet_Company_Kylsant_Empire_Part_6

6. Many sources; see for example the *Scranton* (PA) *Republican*, Nov. 27, 1928, page 2; or the *Huntington* (IN) *Press*, same date, page 1.

7. NASA MODIS website at:
http://earthobservatory.nasa.gov/IOTD/view.php?id=681

8. *Journal of Commerce, The*. Board of Trade inquiry, page 60.

9. Cf. *Dallas Morning News*, November 17, 1928, page unknown. "Crew Talks in *Vestris* Probe." Paraphrased testimony from Tuttle hearings. Cf. *New York Times*, November 17, 1928, page 2. "*Vestris* Men Fail to Explain SOS Delay; Tuttle Charges Witnesses with Evasion on Earlier Message 'We May Need Aid.'" Testimony at the Tuttle hearings.

10. Cf. *Augusta* (GA) *Chronicle*, November 17, 1928, pp. 1 & 2. "Wireless Operator on *Vestris* Says 'Maybe' Captain Radioed Condition Before Disaster." Testimony from the Tuttle hearings. Cf. *New York Times*, November 17, 1928, page 2. "*Vestris* Men Fail to Explain SOS Delay; Tuttle Charges Witnesses with Evasion on Earlier Message 'We May Need Aid.'" Testimony at the Tuttle hearings.

11. *New York Times*, November 29, 1928, page 18. "*Vestris* Inspector Admits False Entry." Testimony at Tuttle hearings. Cf. *Augusta* (GA) *Chronicle*, November 29, 1928, page 2. "Hull Inspector of *Vestris* Claims He Made False Report." Testimony at Tuttle hearings.

12. *Journal of Commerce, The*. Board of Trade inquiry, page 105.

13. *Topeka* [KS] *Plaindealer*, December 7, 1928, page number unknown. "Still Trying to Frame Negro Heroes of *Vestris*." Testimony at the Tuttle hearings.

14. *Augusta* (GA) *Chronicle*, November 18, 1928. Illustration on page 1. (The two illustrations shown here were in a single panel originally.) Credited to "International Illustrated News," which no longer exists.

15. *Augusta Chronicle*, November 18, 1928, page number unknown. "Wind Bitten Mariner Takes Exception to Last Wireless Message Sent by Captain Crane [*sic*]."

16. *New York Times*, November 17, 1928, page 2. "Federal Men Check Clearing of Vessel: Sorensen Denial on Radio." Statement made on radio station WLTH by Capt. Sorensen. A direct quote.

17. *Liverpool Post*, November 22, 1928, page number unknown. "Reporter as Witness."

18. *New York Times*, November 23, 1928, page 2. "Says Capt. Sorensen Assailed Ship's Crew." Paraphrased interview with Fred Puppe about what Captain Sorensen had said aboard the *American Shipper*.

19. — November 18, 1928, page 2, column 3. "Four Leaks Sank *Vestris*, Engineer Says; Chief Officer Didn't Inspect Coal Door Which Crew Says Let in Tons of Water: Contradicts Captain's Message." Johnson's testimony at the Tuttle hearings.

20. *Journal of Commerce, The*. Board of Trade inquiry, pp. 45–46.

21. Ibid. Board of Trade inquiry, pp. 47–48.

22. Goddard, Henry W., United States District Judge for the Southern District of New York, "Decision on the Merits" on the petition of the

Liverpool, Brazil, and River Plate Steam Navigation Co, Ltd, and Lamport & Holt, Ltd., for limitation of liability, as owners of the SS *Vestris*. May 24, 1932. A.M.C. 1932, page 11. At URL: http://patriot.net/~eastlnd2/rj/vestris/Decision.htm

23. *Journal of Commerce, The.* Board of Trade inquiry, pp. 52-53.

24. Ibid. Board of Trade inquiry, page 153.

25. Cf. *Dallas Morning News*, November 17, 1928, page not given (AP). "Suit Filed in *Vestris* Case." Paraphrased AP news report.

26. SOLAS 1929: *International Convention for the Safety of Life at Sea, 1929.* London: His Majesty's Stationery Office, 1932. PDF is at URL: http://www.imo.org/KnowledgeCentre/ReferencesAndArchives/HistoryofSOLAS/Documents/SOLAS%201929%20UK%20Treaty%20Series.pdf

27. Goddard, Henry W., United States District Judge for the Southern District of New York, "Decision on the Merits" on the petition of the Liverpool, Brazil, and River Plate Steam Navigation Co, Ltd, and Lamport & Holt, Ltd., for limitation of liability, as owners of the SS *Vestris*. May 24, 1932. A.M.C. 1932, p. 2.

28. Ibid. pp. 32–33.

29. An optical scan of this letter, which is self-documented as to date, place, and writers, can be found at the following URL: http://patriot.net/~eastlnd2/rj/vestris/23NOV1932.htm

30, *Augusta* (GA) *Chronicle*, May 9, 1929, page 4. "The Truth About the *Vestris*." Op-ed piece quoting extensively from a *Chicago Tribune* op-ed piece. (See next entry.) No author given for the article.

31. *Chicago Tribune*, April 6, 1929, page 14. "The Truth Concerning the *Vestris*." No author listed. The last three of four paragraphs are quoted by the *Augusta Chronicle* article; the first paragraph of the *Augusta Chronicle* article covers essentially the same material as the first paragraph of this article.

Appendix A: Judge Goddard's opinion on the merits in the petition of Lamport & Holt to limit liability

1. Goddard, Henry W., United States District Judge for the Southern District of New York, Admiralty Court, "Decision on the Merits" on the petition of the Liverpool, Brazil, and River Plate Steam Navigation Co, Ltd, and Lamport & Holt, Ltd., for limitation of liability, as owners of the SS *Vestris*. May 24, 1932. A.M.C. 1932. At URL: http://patriot.net/~eastlnd2/rj/vestris/Decision.htm

   This website is a copy of the original judge's decision in the case. It was scanned with optical character reading (OCR) software. Several typical OCR errors were corrected in our text: *the* was misread as *time* in a few places, *miles* was sometimes misread as *rules*, and *but* was misread as *hut* in one place. There were a few other such instances, and there may possibly be some still lurking in the text.

Appendix B: *Vestris* Analysis by K.C. Barnaby

1. Barnaby, Kenneth Cloves, OBE. *Some Ship Disasters and their Causes*. London: Hutchinson Publishing, 1968; New York: A.S. Barnes and Company, 1970. Chapter 6: Accidents and Dangerous Cargoes or Conditions, "Loss of the *Vestris*." Page 167, A.S. Barnes edition. Reprinted by permission of The Random House Group Limited.

Appendix C: Letters written to Judson G. Jackson

1. Letter from William Phipps Adams dated November 30, 1928.

2. Letter from Marion C. Batten, dated December 27, 1928.

3. Letter from Anne DeVore, dated December 12, 1928.

4. Letter from Campbell Kellman, undated.

5. Letter from Dr. Ernst Lehner, dated December 20, 1928.

6. Letter from Captain Frederik Sorensen, dated November 27, 1928.

7. Clippings from a page in the scrapbook of Judson G. Jackson, ca 1928.

All items courtesy of the Jackson Estate and Ramon Jackson.

Other sources of possible interest to our readers

Augusta (GA) *Chronicle*, November 15, 1928. "Human Incidents of Great Tragedy."

—, November 17, 1928. "Forgets Time Message Sent Ship's Agent."

—, November 18, 1928. "How *Vestris*'s Stokers Battled Rising Waters."

—, November 23, 1928. "Water in *Vestris* Hour After Vessel Left, Says Stoker."

—, May 5, 1929. "The Truth About the *Vestris*."

Baltimore News, November 15, 1928. "Six Survivors On The Stand."

New York Times, November 13, 1928. "How Ship Begged Aid Recorded by Radio."

— November 13, 1928. "*Wyoming* Speeds to Scene."

— November 14, 1928. "Cargo Shift Denied; Heavy Seas Blamed."

— November 14, 1928. "Ship Radios Trace Rescue's Progress."

— November 15, 1928. "Officers Describe Rescue on the Radio."

— November 15, 1928. "Passenger Said All Would Have Been Saved."

— November 16, 1928. "Calls SOS Delay by Carey Justified."

— November 16, 1928. "Inspector Insists *Vestris* Was Fit."

— November 16, 1928. "Ship's Cargo Included 4,612 Auto Bodies."

— November 16, 1928. "Tells How Officer Saved Her Twice."

— November 17, 1928. "Federal Men Check Vessel."

— November 23, 1928. "Says Capt. Sorenson [*sic*] Assailed Ship's Crew."

— November 24, 1928. "Firemen Mutinied in Crisis on *Vestris*, Chief Engineer Says."

— December 7, 1928. "*Vestris* Staff Unfit Capt. Jessup Finds."

— December 11, 1928. "254 on *Celtic* Saved After She Hits Rock."

Topeka (Kansas) *Plaindealer*, December 7, 1928. "Still Trying to Frame Negro Heroes of *Vestris*."

Photograph and illustration credits

Cover. Front page of the New York *Daily News* for November 15, 1928. *Daily News* photo archives. Used by permission.

Frontispiece. Photograph by Fred Hanson taken November 12, 1928. New York *Daily News* photo archives. Used by permission.

Page *vi*. Lamport & Holt Line, publicity brochure, ca 1912. Public domain.

Page *xii*. Various thumbnail photos in a montage, printed in the *New York Times* for November 14, 1928. From various sources.

Page *xv*. Lamport & Holt Line, publicity brochure, ca 1912. Public domain.

Page *xvi*. Lamport & Holt Line, brochure cover, ca 1912. Public domain.

Page 5. Inspection certificate for the *Vestris*, reprinted from the *New York Times* for November 16, 1928. Certificate is public domain.

Page 44. Reproduction of drawing by William Adams made November 20, 1928. Public domain.

Page 59. Photo by Fred Hanson, November 12, 1928, New York *Daily News* photo archives. Used by permission.

Page 60. Upper: Photo taken November 12, 1928, by an unknown passenger on the *Vestris*; believed to be public domain. Lower: Photo by Fred Hanson, November 12, 1928, New York *Daily News* photo archives. Used by permission.

Page 61: Photo by Fred Hanson, November 12, 1928. Library of Congress, Prints & Photographs Division, LC-USZ62-31603.

Page 62. Photo by Fred Hanson, November 12, 1928. Published in the *Baltimore News*, November 15, 1928. *Wikipedia* asserts that this photo is now public domain; its copyright expired and was not renewed.

Page 64. Photo distributed by Pacific & Atlantic Photos and published in *The Rockford* (IL) *Daily Register-Gazette* issue for November 15,

1928, on page 6. Copyright expired and was not renewed, so it is now in the public domain.

Page 110. *Rockford* (IL) *Morning Star*, November 18, 1928. Illustration. No credit given.

Page 139. Photograph by naval personnel. From *Wikimedia*, which states that the photograph is in the public domain as a product of the federal government.

Page 140. Photographer and date unknown; photo from the New York *Daily News* photo archives. Used by permission.

Page 141. Artist unknown; illustration from the *Baltimore News*, November 15, 1928. Copyright expired and was not renewed.

Page 142. Photographer unknown, November 14, 1928, at Hampton Roads, Virginia. Photo from the New York *Daily News* photo archives. Used by permission.

Page 143. Photo taken by an unnamed passenger on the *Berlin*, November 13, 1928. Photo from the New York *Daily News* photo archives. Used by permission.

Page 144. Photo by Leonard Detrick, New York *Daily News*, November 24, 1928. From the *Daily News* photo archives. Used by permission.

Page 145. Photo from a Lamport & Holt brochure on the *Vestris*. Public domain.

Page 146: Photo by Acme Newspictures, now defunct. Purchased from Historic Images. Library of Congress says few of these Acme photos were copyrighted, and those copyrights were not renewed.

Page 149. Picture taken by an unnamed photographer at the Tuttle hearings in New York City, November 22, 1928. Its provenance is uncertain at this time. It is on a website as part of the *Time* magazine article "Catastrophe: *Vestris*," which was published November 26, 1928, but Time Inc. denies that it was included in that article. It is probably in the public domain.

Page 150. *Wikimedia Commons*. Photo is in the public domain.

Page 152. This illustration was prepared by author G. David Thayer from an infrared scan of the West Atlantic made by a Moderate Resolution Imaging Spectroradiometer (MODIS) satellite operated by NASA. The original was a false-color image, found at URL: http://earthobservatory.nasa.gov/IOTD/view.php?id=681 Water temperatures, which were represented by the false colors, had to be converted to a grayscale running from black (coldest) to white.

Page 161. Artist's conception of conditions in the stokehold of the *Vestris*. Distributed by International Illustrated News, which no longer exists. Printed in several newspapers, including the *Augusta* (GA) *Chronicle*, November 18, 1928, page 1. First two of four panels.

Page 162. Ibid. Last two of four panels.

Page 177. Upper: Photo taken November 14, 1928, by New York *Daily News* photographer Ed Jackson. Used by permission.
Lower: Picture taken by an unnamed photographer November 14, 1928, on board the *American Shipper*. New York *Daily News* photo archives. Used by permission.

Page 178. Upper: Taken on board the SS *Berlin* by Bob Costa November 14, 1928. New York *Daily News* photo archives. Used by permission.
Lower: Taken November 15, 1928, by Hank Olen. New York *Daily News* photo archives. Used by permission.

Page 270. Photo courtesy of Ramon Jackson.

～❂～

Photo courtesy of Ramon Jackson

The Jackson Family, circa 1925

The families of Ernest Alonzo Jackson and his older brother, Minter Jackson, in front of their home in Abingdon, Virginia. The five sitting between the bannisters in the top line, left to right, are Jannette Beazley Jackson; husband Ernest Alonzo Jackson; Ernest's mother Mary Cloyd Ernest, who married Stephen Alonzo Jackson and was then his widow and remarried to Mr. Davidson; Minter Jackson; and his wife Martha.

In the bottom row, left to right, are Mr. Davidson, Steve Jackson (older brother to Cary), both outside the bannister; Minter Morgan (son of Minter Jackson), Cary Jackson, Lois Jackson (Minter's daughter), Elizabeth Jackson, Virginia Jackson, and Judson Gordon Jackson, (Ramon Jackson's father) with his legs crossed; outside the bannister is Ernest Jackson, the eldest of the children of Ernest and Jannette Jackson. Ernest Alonzo and Jannette Jackson and their youngest child, son Cary, were all lost when the *Vestris* sank. Photo courtesy of Ramon Jackson, whose father, Judson, sent letters to some of the survivors of the wreck. The replies he received are shown in Appendix C.

Index

www.ingramcontent.com/pod-product-compliance
Lightning Source LLC
Chambersburg PA
CBHW060252100426
42742CB00011B/1725